SpringerBriefs in Computer Science

SpringerBriefs present concise summaries of cutting-edge research and practical applications across a wide spectrum of fields. Featuring compact volumes of 50 to 125 pages, the series covers a range of content from professional to academic.

Typical topics might include:

- A timely report of state-of-the art analytical techniques
- A bridge between new research results, as published in journal articles, and a contextual literature review
- A snapshot of a hot or emerging topic
- An in-depth case study or clinical example
- A presentation of core concepts that students must understand in order to make independent contributions

Briefs allow authors to present their ideas and readers to absorb them with minimal time investment. Briefs will be published as part of Springer's eBook collection, with millions of users worldwide. In addition, Briefs will be available for individual print and electronic purchase. Briefs are characterized by fast, global electronic dissemination, standard publishing contracts, easy-to-use manuscript preparation and formatting guidelines, and expedited production schedules. We aim for publication 8–12 weeks after acceptance. Both solicited and unsolicited manuscripts are considered for publication in this series.

More information about this series at http://www.springer.com/series/10028

Mayank Kejriwal

Domain-Specific Knowledge Graph Construction

Mayank Kejriwal
Information Sciences Institute
University of Southern California
Marina del Rey, CA, USA

ISSN 2191-5768   ISSN 2191-5776   (electronic)
SpringerBriefs in Computer Science
ISBN 978-3-030-12374-1   ISBN 978-3-030-12375-8   (eBook)
https://doi.org/10.1007/978-3-030-12375-8

Library of Congress Control Number: 2019931900

This Springer imprint is published by the registered company Springer Nature Switzerland AG.
The registered company address is: Gewerbestrasse 11, 6330 Cham, Switzerland

To the three angels in my life: my mother, my sister, and my niece

Preface

Domain-specific knowledge graphs have emerged as a field unto their own, steadily and perhaps not so slowly. Graphs have been pervasive in AI for a long period of time, dating back to the earliest eras in the field, but automatically representing large quantities of data as graphs is a relatively modern invention. With the advent of the Web, and the need for smarter search engines, both Google and (over a decade later) the Google Knowledge Graph were born. The Google Knowledge Graph has changed the way we interact with search engines, even though we often do not realize it. For example, it is not uncommon anymore for users to not click on a single link when they are searching for something; generally, the search engine itself is able to provide the solution for the problem the user seems to be facing. Organic integration of the traditional search engine with images, news, and videos has only added an element of richness to these interactions.

For all its success, the Google Knowledge Graph (and other similar efforts) was not designed with a specific domain in mind, although Google has rolled out flavors of "domain-specific search" engines (e.g., Google Scholar) every now and then. One would almost be forgiven for thinking that building domain-specific systems, powered by knowledge graphs, for problems such as geopolitical event forecasting, or academic literature mining, is too esoteric to come into its own as an independent, impactful area of study.

What has changed the game and made researchers (and customers) look at domain-specific knowledge graphs as a viable technology is that it has become *easier* to build such knowledge graphs, starting from data collection all the way to the application interface. This was not always the case. Only a few years ago, if I wanted a domain-specific knowledge graph for the e-commerce domain, for example, I would have to assemble a team and build out a system for months before anything remotely viable would emerge. The DARPA Memex program has had an enormous impact in changing this sad state of affairs, by allowing the *democratization* of domain-specific knowledge graph construction. Technologies that emerged from the Memex program combined both classic and state-of-the-art techniques in fields as diverse as information extraction and entity resolution to produce end-to-end systems that could be used by *nontechnical* domain experts to build entire search engines powered by knowledge graphs. A lot of the work that we describe here was rediscovered and utilized in the Memex program to build these end-to-end systems.

Some of the fields that I mentioned above, such as information extraction and entity resolution, are entire areas of study in their own right, with numerous surveys and books individually covering them. Thus, I have had to make some necessary trade-offs in writing this book, and I have chosen to focus on breadth, and comprehensiveness, rather than depth and full academic rigor. In other words, what I attempt to provide in this short work is a comprehensive, practical methodology for constructing domain-specific knowledge graphs using the full range of technology that is available today. I do not shy away from the truism that in many cases, there are no right solutions; one has to deal with compromises. This book tries to detail what these compromises are and when it makes sense for someone wishing to construct domain-specific knowledge graphs to adopt a particular technology or technique.

Since the book is largely based on the findings of multiple communities, there is a lot of credit to go around in conveying the content of each chapter. In some cases, such as IE, I have drawn broadly on widely cited reviews of the field by merging and conveying key elements of both classic and modern surveys, to give the reader a sense of both new developments and established techniques. Because this book is only meant to be a condensed, though hopefully practical and relatively comprehensive, introduction to the field, I have not attempted to provide a rigorous citation for every system or statement. Rather, at key junctures, I have provided pointers to the broader sources that provide a much more comprehensive treatment of related work for the more technically oriented researcher.

I am fairly confident that this book will not provide the last word on this subject. All indicators suggest that research on knowledge graph construction is intensifying, and with increasing synergies between natural language processing, deep learning, knowledge discovery, and semantic web, we will likely see some exciting new work emerge in the years to come. At the time of writing, it is safe to conclude that the field stands at an exciting junction.

Marina del Rey, CA, USA   Mayank Kejriwal
December 2018

Acknowledgments

This book would not be possible without the guidance of, and constant stimulating discussions with, my colleagues and fellow researchers at the Information Sciences Institute. Over the years, we have been jointly funded under multiple projects sponsored by agencies like DARPA and IARPA, covering domains as diverse as geopolitical events, human trafficking, cyberattack prediction, and hybrid forecasting, to only name a few. Many of these involve constructing domain-specific knowledge graphs in support of the final system, where direct or indirect. As such, my time working on some of these projects and collaborating with others on building real applications has led to many of the core findings (and even the structure) in this book.

I also want to thank my students, whose heavy lifting on many of these projects has been at least as valuable to me in learning about knowledge graphs as traditional academic material. I also want to thank the funding agencies themselves, especially DARPA, for sponsoring these students and our work. Ultimately, without their support, this work and its impact would have gone unrealized.

Contents

1 **What Is a Knowledge Graph?** 1
  1.1   Introduction 1
  1.2   Example 1: Academic Domain 4
  1.3   Example 2: Products and Companies 5
  1.4   Example 3: Geopolitical Events 6
  1.5   Conclusion 7

2 **Information Extraction** 9
  2.1   Introduction 9
  2.2   Challenges of IE 10
  2.3   Scope of IE Tasks 11
       2.3.1   Named Entity Recognition 12
       2.3.2   Relation Extraction 22
       2.3.3   Event Extraction 24
       2.3.4   Web IE 26
  2.4   Evaluating IE Performance 29
  2.5   Summary 30

3 **Entity Resolution** 33
  3.1   Introduction 33
  3.2   Challenges and Requirements 34
  3.3   Two-Step Framework 38
       3.3.1   Blocking 39
       3.3.2   Similarity 44
  3.4   Measuring Performance 47
       3.4.1   Measuring Blocking Performance 48
       3.4.2   Measuring Similarity Performance 50
  3.5   Extending the Two-Step Workflow: A Brief Note 51

3.6   Related Work: A Brief Review 51
      3.6.1   Automated ER Solutions 52
      3.6.2   Structural Heterogeneity 55
      3.6.3   Blocking Without Supervision: Where Do We Stand? 56
3.7   Summary 57

4 Advanced Topic: Knowledge Graph Completion 59
4.1   Introduction 59
4.2   Knowledge Graph Embeddings 61
      4.2.1   TransE 63
      4.2.2   TransE Extensions and Alternatives 64
      4.2.3   Limitations and Alternatives 66
      4.2.4   Research Frontiers and Recent Work 66
      4.2.5   Applications of KGEs 72
4.3   Summary 74

5 Ecosystems 75
5.1   Introduction 75
5.2   Web of Linked Data 75
      5.2.1   Linked Data Principles 77
      5.2.2   Technology Stack 78
      5.2.3   Linking Open Data 79
      5.2.4   Example: DBpedia 80
5.3   Google Knowledge Vault 82
5.4   Schema.org 84
5.5   Where is the Future Going? 86

Glossary 89

References 93

Index 103

Acronyms

KG	Knowledge Graph
AI	Artificial intelligence
GKG	Google Knowledge Graph
IRI	Internationalized Resource Identifiers
SW	Semantic Web
URI	Uniform Resource Identifiers
HTML	Hypertext Markup Language
NLP	Natural language processing
IE	Information extraction
KGC	Knowledge graph construction
NER	Named entity recognition
ER	Entity resolution
CRF	Conditional random field
Open IE	Open information extraction
IR	Information retrieval
RNN	Recurrent neural network
LSTM	Long short-term memory
RE	Relation extraction
ACE	Automatic content extraction
MUC	Message Understanding Conference
NE	Named entities
EE	Event extraction
PC	Pairs completeness
PQ	Pairs quality
RR	Reduction ratio
ROC	Receiver operating characteristic
KGE	Knowledge graph embedding
KB	Knowledge base
RDF	Resource description framework
LDA	Latent Dirichlet allocation
RDF	Resource description framework

PSL	Probabilistic soft logic
TKRL	Type-embodied knowledge representation learning
DKRL	Description-embodied knowledge representation learning
LOD	Linking Open Data
GKV	Google Knowledge Vault
KV	Knowledge Vault
OKN	Open Knowledge Network

Chapter 1
What Is a Knowledge Graph?

1.1 Introduction

In recent years, *knowledge graphs* (KGs) have emerged as a major area in Artificial Intelligence (AI) [139]. Graphs have always been pervasive in the broader AI literature, but with the advent of large quantities of data on the Web ('Big Data') and in the broader commercial sphere, there emerged a need to enable machines to 'understand' and make use of this data in some productive analytical way. The inability of machines to truly understand English, and other 'natural' languages like it, with all their irregularities and nuances, has also been largely evident in the (unsuccessful) quest to achieve general AI and commonsense reasoning. Although much progress has been made in all of these domains, it is still very much the case that machines have an easier time processing structured data in the form of graphs, dictionaries and tables than in natural language.

In modern history, Google was among the first big companies to recognize and couple this ability with that of providing richer search capabilities on the Web. In fact, the use of the term 'Knowledge Graph' in recent Computer Science articles, papers and posts, can be traced back to the *Google Knowledge Graph*, which was described in an influential blog post in the early 2010s. The basic motto behind the Google Knowledge Graph was to make search about *things not strings* [164]. In other words, it would allow search to evolve from simple string searching (with all its bells and whistles), to one that involved reasoning about entities, attributes and relationships. The effort can be argued to have been very successful. While the full size and scope of the Google Knowledge Graph is not known, it has grown considerably in size and many search results on Google now involve *knowledge panels* (Fig. 1.1), which are elaborate, yet condensed, information sets about entities that the user might have been searching for. This is in contrast to the previous status quo, which was a list of webpages, ordered by predicted relevance to the user's search query. Beyond Google, other companies have also now started investing in knowledge graphs, and a number of KG-centric startups have emerged in multiple

M. Kejriwal, *Domain-Specific Knowledge Graph Construction*, SpringerBriefs in Computer Science, https://doi.org/10.1007/978-3-030-12375-8_1

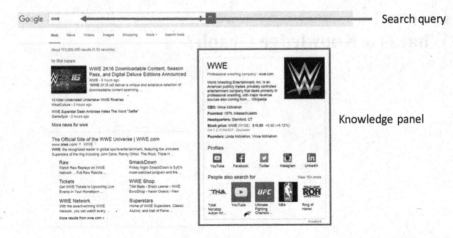

Fig. 1.1 An illustration of a knowledge panel rendered in Google for the search query 'wwe'. At least in part, the panel is powered by KG-centric technologies

countries and continents. There are also applications in non-profit, government and academia. We cover an exciting range of current and growing KG ecosystems in Chap. 5.

Defined *abstractly*, a knowledge graph is a graph-theoretic representation of human knowledge such that it can be ingested *with semantics* by a machine. In other words, it is a way to express 'knowledge' using graphs, in a way that a machine would be able to conduct *reasoning and inference* over this graph to answer queries ('questions') in some meaningful way. However, this definition is not very operational. The simplest *functional definition* of a knowledge graph is that it is a set of *triples*, with each triple intuitively representing an 'assertion'. If the KG was constructed correctly (with 100% accuracy) over a trustworthy data source, we could also think of assertions as *facts*. Formally, a triple is a 3-tuple (h, r, t) where h represents a head entity, t represents a tail entity, and r expresses a relationship between the two entities. Many, though not all, statements in natural language (e.g., English) can be expressed conveniently in this form. Consider, for example, the sentence *Fido the dog stole a bone from Mary's backyard*, which can be expressed as a set of triples[1] {(Fido, is-a, Dog), (Fido, stole, bone_1), (bone_1, is-a, Bone), (bone_1, located-in, yard_1), (yard_1, is-a, Yard), (yard_1, belongs-to, Mary), (Mary, is-a, Person)}.

Why does it make sense to call such a set a 'graph'? For a long time, in fact, it was not conventional to do so and what we are referring to as a knowledge graph here used to be known (and is still known, in many papers) as a knowledge *base*. One of

---

[1]For reasons that will become clear throughout the book, we use identifiers such as bone_1 and yard_1 to refer to *instances* of *concepts* (also called *classes*) such as Bone and Yard. The convention adopted herein is to use capitalized initials for concepts.

the main reasons why knowledge bases slowly morphed into knowledge graphs can be attributed to the influence and success of the Google Knowledge Graph. However, there was also pervasive influence from both the knowledge discovery, and Semantic Web, communities, both of which have always been closely associated with graph-theoretic innovations. For a large part of this millennium, the database community was also studying graph databases, algorithms and data structures in detail.

This fascination (both industrial and academic) with graphs aside, there was another good reason to think of knowledge bases as graphs. First, if one takes the step of visualizing the first and third elements (i.e. h and t) of a triple as *nodes*, and the second element r as a labeled, directed edge pointing from h (the *head* entity) to t (either a *tail* entity or an *attribute*), an intuitive data model emerges (Fig. 1.2). In fact, many people would find it easier to draw the kind of diagram shown in Fig. 1.2 (with a few examples for guidance) than thinking carefully about sets of triples. In a certain sense, the KG can be said to serve as a lingua franca between machines and humans, in that it is structured enough for machines to process and ingest with semantics, but is intuitive enough for humans to make sense of, at least if represented and drawn using common-sense mnemonics. In fact, the Freebase knowledge graph, and more recently, Wikidata, allow the crowdsourced acquisition of such structured knowledge, as opposed to Wikipedia, where the crowdsourced knowledge is acquired mostly in natural language.

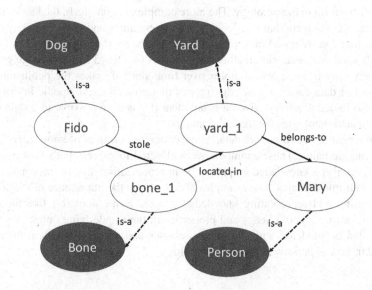

Fig. 1.2 The Knowledge Graph (KG) representation of the information expressed in the Fido the dog example. Filled ovals (i.e. concepts) are parts of the ontology, while the unfilled ovals are part of the KG itself (the instances) is-a relationships (dashed edges) mediate between instances (in the KG) and concepts (in the ontology). Other relationships are defined in the ontology, but used in the KG. Constraints on how the relationships may be used are considered part of the ontology, typically defined using formal declarations

Although the simple definition (which we shall refer to as the 'knowledge base' or KB-definition, where relevant) has many advantages, not the least of which is its simplicity and ease of reading into, and serializing from, machine learning and other data analytics programs, it is also unsatisfactory for certain applications. Just like we do not want a database to have an 'open' schema, we do not always want a knowledge graph to be unconstrained in terms of the data it contains, and the ways in which that data is modeled. This leads to the notion of an *ontology*, which (put simply) defines (and imposes constraints) on the concepts and relationships that are permissible in a KG. For example, considering the earlier example of Fido the dog, it is clear that the ontology contains concepts such as *Dog*, *Bone*, *Yard* and *Person* and defines relationships as well. An example of a defined constraint is that the 'belongs-to' relation must have a *Person* instance (e.g., Mary) as its *target*. Considering the example in Fig. 1.2, the is-a relationship mediates between the KG, which contains instances, and the ontology, which contains concepts. Although the example makes it look straightforward, it can sometimes become a point of contention as to what is a concept and what is an instance in the real world.

Beyond the Google KG, most KGs are domain-specific and have some kind of underlying ontology. This is because there is typically no 'one-size fits all' schema or ontology that is well-suited for solving all problems or answering all queries. Deciding what makes for a good ontology is a controversial topic that is outside the scope of this book. However, once an ontology is given, the expectation is that the KG will conform to the ontology. The more complex the ontology, the harder it is to make the KG conform, but the stronger are its semantics and the complexity of the queries that can be posed on the knowledge graph. As the community has moved towards statistical, data-rich methods, ontologies (designed for knowledge graphs) have generally become less complex over time since it makes the publishing and checking of data easier. Knowledge graphs that contain encyclopedic information have also fueled this trend, since it is not clear if it is even possible to design deep, sophisticated ontologies for 'broad' domains.

In the next few sections, we detail some concrete examples of knowledge graphs in various domains. These examples were selected to express both how intuitive, and expressive, a knowledge graph can be in representing diverse information sets across multiple domains. The examples also illustrate the importance of the domain in modeling and representing knowledge graphs. Some domains, like the event domain, require lots of classes and properties in their underlying ontologies while others can be modeled with only a few classes and properties. Often, there is a choice in how expressive to make the ontology.

1.2 Example 1: Academic Domain

As our first example of a domain-specific KG, let us consider the academic *publication* domain (Fig. 1.3). The two purple nodes in the center of the KG represent different publications, named mnemonically by their publication titles.

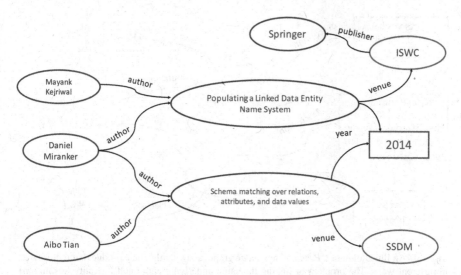

Fig. 1.3 An illustration of a *Publication* knowledge graph, showing two different publications sharing authors. Rectangles are typically used diagrammatically to represent *literals* while oval nodes represent *resources* or *entities*

Some important details concerning the publications are also shown, including their authors, dates of publication and venues.

Despite its simplicity, the KG in Fig. 1.3 illustrates some of the expressiveness in *representation*, an issue that becomes extremely important in communities such as the Semantic Web (SW). The oval nodes in the figure represent *entities* or *resources*, and are generally referred to (in the SW community) as Internationalized Resource Identifiers (IRIs), a generalized form of *Uniform* Resource Identifiers (URIs). In this book, we do not define these concepts, since they are community-specific, but focus more on the overall distinction between entities and literals (also known as *attributes*). Entities can have relationships with other entities (such as between authors and their publications) or attributes (such as the year of a publication). The distinction can be expressed by the fact that in a triple (h, r, t), t is either a literal (for the latter) or an entity (for the former). Note that h is always an entity.

1.3   Example 2: Products and Companies

In the second example, inspired by the *products and e-commerce* domain, we expand upon the notions presented in academic domain. Once again, we see the distinction between literals and entities, but as illustrated in Fig. 1.4, there are numerous degrees of freedom even when modeling the most basic structures in KGs. In this case, we see the same product, represented and modeled in two different ways. The choice of modeling can have implications both for upstream tasks (such as information

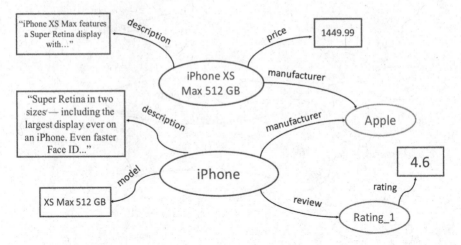

Fig. 1.4 An illustration of a *Product* knowledge graph, showing the same product but represented in different ways. The problem of linking the same underlying entity nodes (Entity Resolution) will be covered in detail in Chap. 3

extraction) and downstream tasks, such as entity resolution and querying, that occur after the initial KG has been extracted and stored. We also see that the availability of information can vary, usually depending on the source from which the KG nodes were extracted to begin with. Also, because the two *product mentions* have not been resolved into a single underlying entity, it is not straightforward to compute an *aggregation* (e.g., the number of unique products) over such KGs and expect reasonable or correct answers. Because it is often the case that the same entity is extracted independently from multiple raw sources, one has to perform *Entity Resolution* on the extracted KG. We cover this step in more detail in Chap. 3.

1.4   Example 3: Geopolitical Events

Finally, we consider the most complex, and cutting-edge, example of a geopolitical event KG. In addition to the usual artifices that we saw before (entities vs. literals etc.), the graph illustrates how 'second-order' entities like events can be represented in a KG. Events are second-order because they have first-order entities like locations and times as their *arguments*; in turn, these first-order entities have attributes describing them further (Fig. 1.5). Events can also directly have attributes, and similar to first-order entities, have relationships between themselves. The notion of what separates first-order from second-order entities is not completely well-understood and is more of a semantic rather than a syntactic issue. In practice, the difference is very real. Extracting and resolving events, for example, have become areas of research in their own right, and performance on them continues to be poor

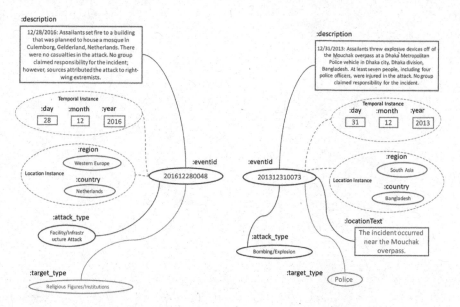

Fig. 1.5 An illustration of an *Event* knowledge graph, showing two disparate geopolitical events

in comparison to performance on extracting and resolving first-order entities like persons and locations. A good example of an event knowledge graph is GDELT [101].

1.5   Conclusion

Knowledge graphs have become a popular data representation that sits at the intersection of knowledge discovery, data mining, Semantic Web and Natural Language Processing. Each of these communities has had dealings with knowledge graphs and their applications. In fact the term is so broad that there is no real 'survey' of knowledge graphs, making it difficult to attribute the invention to a specific paper or author. Generally, the focus is on 'generic' knowledge graphs without too much emphasis on the domain, and domain-specific constraints, that girds the construction and representation of the knowledge graph. An increasing amount of evidence suggests that there is no one size fits all model for knowledge graph construction and inference that can be used across all domains; rather, special domain-specific techniques must be used to obtain state-of-the-art performance. In the rest of this book, we cover domain-specific knowledge graph construction in detail. Although the area is continuing to evolve, some trends have been established and are built on prior research developed over multiple decades. At the time of writing, knowledge graph-powered applications continue to proliferate (Chap. 5).

Chapter 2
Information Extraction

2.1 Introduction

Information extraction (IE) is a fundamental component in any knowledge graph construction pipeline, whether domain-specific or generic [111, 119]. As the name implicitly suggests, the goal of an IE system is to extract *useful* information from 'raw' data, usually text or webpages. Useful information has many dimensions, but the most important dimension for computational purposes is that the information can be queried, and reasoned about, by machines. One of the reasons why IE was identified as an early problem in communities like natural language processing was because machines are not good at understanding natural languages like English or French due to problems like subtlety, ambiguity and irregularity. Even today, despite rapid advances, machines still cannot read and understand English nearly as well as humans. Thus, an early goal of IE was to extract key pieces of information, such as entities, relations, events and attributes from natural language text. The advent of the Web only made the problem more interesting, since webpages are visually intuitive (if rendered in an compatible browser), but in their raw HTML versions, contain many interesting 'markup' elements like tables, lists, links, images and even dynamic elements like Javascript programs [36].

In Chap. 1, we implicitly demonstrated the results of 'perfect' information extraction when we converted an English sentence (about Fido the dog) to 'equivalent' knowledge graph triples. To take another example, a sentence such as 'Vladimir Putin, President of Russia, attended the G20 meeting', can be represented by the following set of triples: {(Vladimir Putin, presidentOf, Russia), (Russia, is-a, Country), (Vladimir Putin, is-a, Person), (G20, is-a, Geopolitical Meeting), (Vladimir Putin, attended, G20)}.

Recall that we referred to elements like 'Vladimir Putin' as entities, while elements like 'presidentOf' are generally referred to as relations, relationships or properties. We also introduced the notion of literals (also called attributes or slot fillers depending on both the context and community) in the previous chapter; e.g.,

M. Kejriwal, *Domain-Specific Knowledge Graph Construction*, SpringerBriefs in Computer Science, https://doi.org/10.1007/978-3-030-12375-8_2

if we had a triple such as (Vladimir Putin, DOB, "10/07/1952"), the date of birth would be a triple. Generally, numbers and strings are understood to be literals.

The example above notwithstanding, the term 'IE' is far too broad to be useful in practice, since there is no one IE system that can extract all possible entities and relationships from a ('completely triplify') given English (or other natural language) sentence. Generally, IE systems are constrained by an underlying ontology, just like an actual knowledge graph, although in recent years, the concept of Open IE (which is open-world and not constrained by an ontology) has been gaining some traction [10].

In this chapter, we take a practical view of IE based on established literature. Over the last few decades, IE has been heavily researched and many techniques are currently in use in the community, including classic rule-based techniques, more modern sequence labeling techniques such as Conditional Random Fields (CRFs) [100] but also more cutting-edge techniques such as deep neural networks [47]. We begin by describing why IE is so challenging, and why it continues to be the most important component in building high quality KGs from scratch. Next, we survey IE from a range of functional perspectives.

2.2   Challenges of IE

IE has been explored in the AI community for several decades now [48, 161]. That the problem is still actively researched and has not been solved yet is a testament both to its difficulty and its relevance. In this section, we explore some predominant challenges that prevent IE systems from reaching arbitrarily high quality for many real-world datasets.

First, state-of-the-art IE systems tend to be based on supervised machine learning, a class of techniques whose success is predicated on having access to labeled training data. Labeling data from scratch is an expensive and time-consuming endeavor that cannot be effectively scaled. In some cases, the labeled data itself (as opposed to the algorithms trained on the data) is the basis for a system's competitive advantage, and is closely guarded, especially in the commercial sphere. However, due to efforts like the Message Understanding Conferences (MUCs) [68], the research community as a whole has come a long way in developing a robust set of benchmarks both for training and evaluating IE algorithms. However, even within this body of labeled data, some IE tasks (like named entity recognition) are much better supported than others. As novel IE tasks and data emerge, such as joint text-video extraction and event extraction, the utility of previously labeled datasets become less clear [175].

Second, for domain-specific KGC, IE presents some additional challenges. Like with so many techniques that use machine learning and are optimized for a 'generic' (or open-world) domain such as the encyclopedic world covered by Wikipedia or the Google news corpus, problems especially arise in domains that are different from these common corpora and require special techniques for maximal performance.

The question that arises is: how does one build reasonably high-quality IE systems *without* access to a lot of labeled data? On a related note, how does one make *judicious* use of unlabeled data? What *kinds* of supervision are possible besides mundane labeling of lots of data? These questions have been explored in the research community for quite some time, and we cover some techniques in this chapter.

Third, the format and *heterogeneity* of the raw data is very important and can also be a challenge when transitioning results across communities or research groups. Are we extracting information from HTML webpages or from a plain text file? Within an HTML file, do we have a lot of tables, markup and even javascript? Isolating the relevant information from the page before running an IE system over it can itself be a challenge. We describe in a later section how wrapper induction techniques can be used to extract meaningful information from webpages [99].

2.3 Scope of IE Tasks

Because IE is such a diverse problem, any review must necessarily *scope* the problem. In some cases, the tasks are different enough that they get separated by community. Multiple surveys and reviews also tend to take this view; see for example [161] and [111]. Where possible, we attempt to follow a similar flow as others that have extensively surveyed individually components of this chapter. For example, much of the work on NER described in this chapter is closely inspired by the survey of Nadeau and Sakine [121]. For example, traditionally, it has been the case that Web IE was treated very differently from NLP-centric IE like extracting named entities, relations and events from text. Although these distinctions still remain, their significance has diminished, in part because even within each community a wide range of techniques and methodologies have flourished. For this reason, we do not separate Web IE and NLP-centric IE in this chapter, but consider them as different IE *tasks*. Among the different tasks below, the first three (Named Entity Recognition, Relation Extraction and Event Extraction) have been overwhelmingly researched by the NLP community. The last task, Web IE, has witnessed more research attention in the overall AI literature, with wrappers emerging as a dominant technique even in the early days of the Web. Machine learning has been used extensively for all the different IE tasks described below. Interesting combinations are also possible. For example, given a corpus of text-heavy webpages (e.g., blog articles scraped from an online portal), one may initially run a wrapper or Web IE system to strip out the HTML boilerplate, and obtain the underlying text. Next, a sufficiently trained and tuned IE for tasks such as NER and relation extraction may be applied.

In fact, many such combinations are possible, and the flexibility, architecture and quality of an overall IE pipeline depends significantly both on the experience and imagination of an application designer. The tasks below are not expected to be mutually exclusive, and some are mutually *reinforcing*. We illustrate a particularly important case of this when we describe joint event-entity extraction.

2.3.1 Named Entity Recognition

Named Entity Recognition (NER) is often the first line of attack in a given IE problem domain [121]. Given a document and a set $T = \{t_1, \ldots, t_n\}$ of n entity types (defined in an ontology, as previously covered), a NER system generally returns a set of *extracted mentions*, where each mention may be expressed in the form *(t, start-offset, end-offset)*, with $t \in T$. Consider the sentence '**Tom Cruise** shot the latest Mission Impossible movie in **Dubai**', where the extracted mentions are in bold. Given the type set {Person, Location}, the mention 'Tom Cruise' is clearly of type Person, while 'Dubai' is of type Location. It is precisely because the mentions are typed that they are often referred to as *named entities*. However, the term is also a misnomer, since mentions that refer to the same underlying entity need to be resolved to a single named entity. This process, called Entity Resolution (ER) in the graph-theoretic and broader KG community [66], will be covered in the next chapter. In a similar vein, the step of 'coreference resolution' is also often undertaken in the NLP community to leverage text and syntax features (such as part-of-speech tagging and Wikipedia linking [73, 173]) to link different mentions, including pronouns, to each other [122].

As the example sentence above illustrated intuitively, full named-entity recognition can be broken down into two distinct problems: detection of named entities, and classification of the named entities by the ontological type (i.e. concept). The first phase is akin to *segmentation* (which falls in the same category as other shallow parsing tasks e.g., *chunking* in the NLP community [97]). More generally, named entities are defined to be contiguous *non-nested* spans of tokens, so that 'President Barack Obama' is a single named entity, disregarding the fact that inside this name, the substring 'Barack Obama' is itself a valid named entity. This can sometimes lead to interesting problems for evaluating IE, especially at scale. Whether 'Barack Obama' or even 'Obama' would be tagged as correct instances of detection, as opposed to the more complete and unambiguous 'President Barack Obama' depends on the application and the way in which the evaluation is set up.

The second phase would require assigning a concept from T to the detected named entity. Mis-typed named entities would lead to errors in evaluation, as we describe towards the end of the chapter. Intuitively, we do not want to tag 'Tom Cruise' as a location rather than a person.

In practice, the situation can become rapidly more complex, since, in the definition of T above, there is no reason to restrict it to *just* a set of types. In fact, in a more complex task, such as event extraction, it is more reasonable to allow T to represent the *ontology* itself. Some event ontologies, such as CAMEO [180], can be very hierarchical and contain fine-grained classes like 'reject accusation', 'deny responsibility', 'express accord' and 'appeal for diplomatic cooperation'.

With such hierarchical ontologies, one would ideally want to tag a detected mention with the most fine-grained type, which is more difficult than distinguishing between coarse-grained concepts like Location and Person. For example, one could imagine an ontology that has fine-grained types such as 'Politician', 'Businessman', 'Celebrity', many of which would be linked to a super-type such as 'Person'. Given such an ontology, would it be incorrect to tag 'Tom Cruise' with type 'Businessman' or would 'Celebrity' be the only correct answer? How do we penalize a system which tags 'Tom Cruise' as 'Person' vs. a more sophisticated system that tags the mention as 'Celebrity'? In general, evaluating the outputs of NER with respect to such an ontology is itself a complex issue, and in practice, depends on the application. We do not consider the issue in detail here. In the rest of the section on NER, we restrict our attention to the most popular version of the problem, which assumes a single non-hierarchical set T of terms as the ontology. Later, we briefly discuss an exception to the availability of an ontology (Open IE [10]).

Before describing the classes of techniques that are generally used for NER, we note that the question of how to precisely *define* a named entity is important philosophically, but is almost always evident from the application context. A functional definition is that a named entity is simply a span of text that can be typed according to one or more classes in a pre-specified ontology. The last word on whether a named entity is correct or not mechanistically depends on the gold-standard, which can itself occasionally contain missing and wrong entries, both due to human error or because of annotator disagreements.

2.3.1.1   Supervised Methods

The current dominant technique for addressing the Named Entity Recognition problem is supervised machine learning. Such techniques have classically included Hidden Markov Models, Support Vector Machines (SVMs) and Conditional Random Fields (CRFs) [78, 100, 191]. Although CRFs have emerged as the popular technique due to superior performance, in general, all of these machine learning methods can be used to construct a system that is trained using a large annotated corpus. In essence, such a system creates disambiguation rules based on discriminative features. An obvious baseline statistical learning method that is often proposed consists of tagging words of a test corpus when they are annotated as entities in the training corpus. The performance of the baseline system depends on the *vocabulary transfer*, which is the proportion of words, without repetitions, appearing in both training and testing corpus. As reported in a review, in a vocabulary transfer study that was conducted on the MUC-6 training data, it was found that vocabulary transfer accounted for 21% of performance, with about 42% accounting for locations, 17% for organizations and 13% for person names [134]. What these numbers illustrate is that achieving *generalization* is an important requirement, but that considerable variance exists across common ontological types. Vocabulary transfer is also a good signal of the recall (number of entities identified over the total number of entities) of the baseline system but (on the flip-side) may be more pessimistic than it appears since some entities are more frequent than others.

2.3.1.2 Semi-supervised Methods

Unlike supervised learning, semi-supervised learning (sometimes also known as *weakly supervised*) is designed to significantly reduce labeling effort without removing the human in the loop altogether (unlike *unsupervised learning*, which is covered subsequently) [120, 149]. An example of an influential semi-supervised NER technique is 'bootstrapping', which generally requires a set of seeds for initiating the learning process. For example, a system aimed at recognizing case citations in legal documents might ask the user to provide a small number of example citations. Then the system searches for sentences that contain these names and tries to identify some contextual clues (e.g., surrounding words, or their word embedding representations) common to the provided examples. Subsequently, the system tries to find other instances of citations that appear in similar contexts. This learning process is reapplied to the newly found examples, to discover more new relevant contexts. By iterating and repeating this process, a large number of case citations and contexts will eventually be discovered. Although supervised learning continues to be state-of-the-art, some experiments in semi-supervised NER have yielded performance rivaling baseline supervised approaches. Recently, these methods have become even more important in the context of extracting named entities from Twitter with minimal supervision [107, 156]. Thus, specific techniques are worth considering. We provide a brief review of some influential methods below

Mutual bootstrapping Mutual bootstrapping, first introduced in [155] as multi-level bootstrapping, consists of *growing* a set of entities and contexts by starting with a handful of seed entity examples of a given type (e.g., Sharon Stone and Sylvester Stallone are entities of type Actor) and *accumulating* the patterns found around these seeds in a sufficiently large corpus. Contexts, which are like linguistic patterns, (e.g., starred in X, is the star of X) are also ranked and used to find new examples to achieve some semblance of generalization. The authors of the original paper note that performance can rapidly deteriorate as noise starts to creep into either the entity list or pattern list. Although the overall empirical results were mixed, the idea behind mutual bootstrapping proved to be highly influential and has been considered as a prominent approach since, with multiple proposed variants. For example, one variant is to use syntactic relations (subject-object pairs) to describe context around the entities, rather than simple linguistic patterns. Furthermore, rather than relying on human-generated seeds, the process could be automated by relying on existing NER system outputs instead. In yet another variant, distributional similarity has been leveraged to generate synonyms or words which are members of the same semantic class [34], which allows for more robust *pattern generalization* (which, broadly speaking, is what much of NLP is about [109]). For instance, for the pattern X holds his meeting on a Monday, synonyms for Monday would be

{Tuesday, Wednesday, Tue., ... }, thereby enabling the induction of more novel and generalizable patterns. Several authors have demonstrated that, by applying this technique to large corpora (hundreds of millions of webpages) and starting only from a seed of 10 examples facts, it is possible to generate one million facts with a precision of about 88%, an impressive performance metric.

How should one go about selecting the unlabeled data on which such mutual bootstrapping methods (or their variants) can be applied? One way is to rely not just on large arbitrary collection of documents, but to select documents using information retrieval-like relevance measures [160]. Furthermore, selection of specific contexts that are rich in proper names and coreferences bring the best results in their experiments. In general, the data selection problem is still not completely solved, especially given the almost limitless amounts of data (on any domain) now available on the Web. This problem has emerged into its own, and goes by various names, including intelligent crawling and domain discovery. For a treatment on the subject, see [133].

2.3.1.3   Unsupervised Methods

Clustering is the quintessential unsupervised machine learning approach [79]. Clustering of entities can rely on the similarity of context, lexical resources (e.g., WordNet), and even lexical patterns and statistics computed on a sufficiently large. Some specific approaches are described below.

As one form of unsupervised clustering, [4] studies the problem of labeling an input word with an appropriate named entity type from WordNet, the primary approach being to assign a *topic signature* to each WordNet synset by merely listing words that frequently co-occur with it in a large corpus. Then, given an input word in a given document, the word context (words appearing in a fixed-size window around the input word) is compared to type signatures and classified under the most similar one.

In a similar vein, a method similar to the identification of hyponyms/hypernyms described originally in [74] was proposed to identify potential hypernyms of *sequences* of *capitalized* words appearing in a document. Other variants have been proposed as well.

The observation that named entities often appear together in several news articles, whereas common nouns do not, has also been leveraged as 'background' knowledge for unsupervised NER model building. For example, a strong correlation was found between being a named entity and appearing contemporaneously in multiple news sources, which has allowed the identification of rare named entities in an unsupervised manner and can also be effectively combined with other NER methods.

The authors in [59] proposed and Information Retrieval (PMI-IR) as a feature to assess whether a named entity can be classified under a given type. PMI-IR was originally designed to measure the dependence between two expressions using web queries. A high PMI-IR meant that expressions tended to co-occur. The technique was leveraged for unsupervised NER by creating features for each candidate entity (e.g., Sharon Stone) and a large number of automatically generated discriminator phrases like 'is a movie star', 'starred in', 'won the Golden Globe' etc.

2.3.1.4 Features

Even the brief descriptions of approaches listed above for supervised, unsupervised and semi-supervised NER show that *features* are extremely important for good performance, and more often than not, can prove to be decisive factors in the quality of a system. Furthermore, in domain-specific KGC systems, whether for biomedicine, chemistry or even space, features can play an even more important role than usual [158, 162].

Features are characteristic attributes of words or other (such as segmented noun phrases) that are especially important for robust and good performance of machine learning classifiers. As an example, consider a quantitative feature that counts the number of characters in a word. Features represent abstract properties of a unit that assist in the generalization capabilities of machine learning capabilities. Boolean features return either true or false (e.g., whether a word is capitalized or not), while quantitative features return numbers, whether real-valued or discrete. Features can also be *nominal* or *categorical*[1] e.g., an 'identity-lowercase' feature would simply return the lower-cased version of the word as its output.

Although features have come to be associated closely with machine learning, expert and rule systems also make use of features. We may choose to institute a rule in our NER that says that if (1) the capitalization feature returns true, and (2) the nominal feature 'inc' or 'corp' (and other variants that we decide a priori) is detected, then the 'slot' should be output by the IE system with type Organization. Although rules work well for regular sentences and constrained grammars and styles, real systems tend to be much more complex and rules have to be *induced* using learning techniques or have to be superseded (as state-of-the-work often does) by machine learning-based sequence-labeling using cutting-edge deep neural networks like Recurrent Neural Networks (RNNs), which use units such as Long-Short Term Memory (LSTMs) units for advanced state-of-the-art sequence labeling, or more classic Conditional Random Fields (CRFs).

Features most often used for NER can be categorized along three different axes, as described in the survey by [121]: *word-level*, *list lookup*, and *document and corpus*. We describe these below, followed by notes on some recent advancements in *representation learning* that have fundamentally influenced feature engineering [118].

---

[1]One could also imagine *ordinal* features, which would be helpful in tasks such as recommendation, but these tend to be less common in IE.

Word-level features Word-level features tend to describe the '' of a word, including such aspects as word case, punctuation, special characters and numerical value. It is especially important not to underestimate the usefulness of digits when defining word features. Digits can be used to express such information as dates, percentages and intervals, but each of these tend to be expressed using specific patterns. For example, a pattern such as 'aa?' where '?' expresses a placeholder for a sequence of consecutive numerical characters, could be used to express e.g., a flight number (for American Airlines). Similarly, one can define patterns such that two-digit and four-digit numbers can stand for years, one and two digits may stand for a day or a month and so on [188].

Morphological features, heavily featured in any sufficiently robust NLP pipeline, including for tasks extending beyond (or preceding) NER, relate to elements such as words affixes and roots. Words ending in 'ist' (physicist, radiologist) could be used, with some probability, to detect Person entities that are professionals. Words ending in affixes like 'ish' could indicate nationalities, or even languages e.g., Danish [22]. However, one must be careful, since there are exceptions. For example, it would not be correct to tag the word 'apologist' as a Person entity with a professional background. There is an obvious multi-lingual element to these features as well [13]. For these features to provide value, they must be combined with other features in sufficiently robust learning algorithms. Given a large background corpus, statistical methods can also be used to discover relevant morphological features that may have been overlooked by a feature engineer.

As the examples and feature categories above suggest, features can also be extracted by applying functions over words. Collins and Singer [46] provides an early example where the authors construct a feature function that operates by isolating and concatenating non-alphabetical characters from an input word. For example, given the word 'T.J.Max', the output of the feature function would be '..'. Many candidates like this function now exist in the hundreds of NLP papers describing both statistical and rule-based systems for a variety of problems, including NLP. One of the more popular and recurring features is character n-grams, introduced by [138], although variants of the idea had been around for a while, especially in the information retrieval (IR) community. This feature is important enough that we describe the procedure for extracting it below. It is also slightly unusual in that the output is not a single value but a set of values. In popular variants, the set may also be a multi-set or a 'bag' i.e. it may contain duplicate elements.

Character n-gram Feature Function Let us assume a character n-gram feature function, where n (≥ 1) is known in advance, and a special character #. Given a character sequence (which could be a single word, but can also be a multi-word sequence; the only general constraint is that the input should be a sequence and that the sequence should not include the special character) $[c_1, \ldots, c_m]$ containing $m \geq 1$ characters, the function would first pre-pend and append $n-1$ #'s to the sequence, thereby creating a new sequence of length $m + 2(n - 1)$. Next, a window of size n is slid over the sequence from beginning to end, and the sub-sequence contained within this window is placed inside the output set.

As an example, suppose we wanted to extract 3-grams (called tri-grams) given the chunk 'President Obama'. First, we would append and pre-pend $n-1$ #'s (##) to the sequence, yielding the new sequence (represented as a string for convenience) '##President Obama##'. Now, a window of size 3 is slid over the entire sequence to yield the set {##P, #Pr, Pre,. . . ,ama,ma#,a##}.

Despite its simplicity, tri-grams and bi-grams have some advantages that make them robust to artifacts like spelling errors. Also, characters at the beginning and end of the chunk are overweighted compared to characters in the middle. This is in recognition of the fact that beginnings and endings tend to be important when comparing named entities. Character n-grams are also complementary to another line of important techniques proposed in the information retrieval community decades ago for robust comparison of 'bags of words', namely token-based and set-based similarity measures like cosine similarity on tf-idf vectors.

Pattern features were introduced by [45] and then used by others [44]. The idea is to map words onto a small set of patterns over character types. The goal is to achieve robustness and generalization by performing such a mapping. For instance, a pattern feature might map all uppercase letters to 'A', all lowercase letters to 'a', and all punctuation to '.'. With such a representation, seemingly different words begin to look similar e.g., 'ABC Corp' and 'NBC News' would both map to the same pattern using just the simple rules described above. It is possible to have pattern features that are much more sophisticated, and one could also induce relevant type-customized patterns from a background corpus of named entities. Such sophistication can be a double-edged sword in its own way. For example, if the background corpus is too general, performance of such pattern features may not be well suited or applicable to more domain-specific problems. On the other hand, if significant effort is expended to develop such corpora or patterns for domain-specific problems, then it may not be easy to extend to new domains. Robustness should also be given priority; small changes in inputs or background corpora should not strongly affect results on unknown or unseen (but still same-domain) datasets.

List-lookup Features In knowledge graph construction systems that often rely on background knowledge, such as knowledge derived from public sources like Wikipedia, DBpedia and YAGO [8, 167], list lookup features are extremely important. In traditional work, terminology can be varied, with *gazetteer*, *lexicon* and *dictionary* all used interchangeably with *list*. Lists are useful not just for named entity detection (the probability that Buenos Aires is a named entity rather than Buenos and Aires separately, increases significantly when we observe a Buenos Aires in an external lexicon), but more importantly, for entity typing. In the case of Buenos Aires, it is unlikely we will ever observe the named entity in a context other than as having type 'Location' or 'City' (if finer-grained typing is supported). In other situations, there can be considerable ambiguity. In earlier examples, we described how locations can sometimes be confused with geopolitical entities and vice versa. Because of word polysemy [151], additional problems arise. In the case of domain-specific KGC, especially for non-English scenarios, good dictionaries may not even exist. It is also important to realize a fundamental limitation of knowledge bases like Wikipedia, which generally contain pages on 'well-known' entities. In many domains, it is the long tail of less well-known entities, relations and events that is of interest to analysts and users.

Sometimes, however, lists exist in 'plain sight'. Common nouns listed in a dictionary are useful, for instance, in the disambiguation of capitalized words in ambiguous positions (e.g., sentence beginning). Mikheev et al. [117] reports that from 2677 words in ambiguous position in a given corpus, a general dictionary lookup allows identifying 1841 common nouns out of 1851 (99.4%) while only discarding 171 named entities out of 826 (20.7%). In other words, 20.7% of named entities are ambiguous with common nouns, in that corpus.

Many authors propose to recognize organizations by identifying words that are frequently used in their names, such as 'Inc', 'Corp', 'Associates' or even 'Telecom'. Most approaches implicitly require candidate words to exactly match at least one element of a pre-existing list. However, we may want to allow some flexibility in the match conditions. We describe some possibilities below.

First, words can be stemmed (stripping off both inflectional and derivational suffixes) or lemmatized (normalizing for inflections only) before they are matched [43]. For instance, if a list of cue words contains 'subsidiary', the inflected form 'subsidiaries' will be considered as a successful match. For some languages [80], diacritics can be replaced by their canonical equivalent (e.g., naïve would be replaced by naive).

Second, candidate words can be fuzzily matched against a reference list using techniques like thresholded edit-distance [171] or Jaro-Winkler [44]. This allows capturing small lexical variations in words that are not necessarily derivational or inflectional. For instance, John could match Johnny because the edit-distance between the two words is sufficiently small. In fact, Jaro-Winkler's metric was specifically designed to match proper names following the observation that the first letters tend to be correct while name ending often varies. Other string similarities exist for specific matching tasks.

Third, phonetic algorithms like Soundex or Metaphone can be used to match against a reference list [146], since such algorithms normalizes candidate words to a phonetic code such that names which sound very similar map to the same code e.g., Jon and John would map to the same code. Soundex, which is the oldest and best known of the phonetic algorithms, produces a code which is a combination of the first letter of a word plus a three digit code that represents its phonetic sound.

Document and Corpus Features In the most general case, document features are defined over both document *content* and *structure*. This section describes some features that go beyond simple single and multi-word expression features to include meta-information and statistics about documents and corpora.

In an early work on document-centric features, several authors extract features from documents simply by identifying words that appear *both* in uppercased and lowercased form in a single document [117, 152, 169]. Those words are hypothesized to be common nouns appearing dually in ambiguous (e.g., the beginning of a sentence) and unambiguous positions.

Another feature, which is recognizing multiple occurrences of a unique entity in a document, dates back to research in the field that started in the early 1990s [115]. Determining multiple aliases[2] of an entity is a variant of the famous coreference resolution problem [122], which is still not solved. In early research, deriving features from coreference is mainly done by exploiting the *context* of every occurrence (e.g., Obama signed a treaty, Obama said that taxes will not be raised, Obama declared a truce...). Deriving features from aliases is mainly done by leveraging the union of alias words (Sir, Elton, E., John).

Finding coreferences and aliases in a text can be reduced to the same problem of finding *all* occurrences of an entity in a document, a complex endeavor. In a well-known domain-specific example, for instance, the authors in [64] use 31 heuristic rules to match multiple occurrences of company names. As an example of a heuristic, two multi-word expressions match if one is the initial subsequence of the other. Cross-document coreference resolution is even more complicated, on which research continues to this date. Both supervised and unsupervised approaches have been proposed and compared [103]. One proposal is to use word-level features engineered to handle equivalences (e.g., dr. is equivalent to doctor), with relational features encoding the relative order of tokens between two occurrences.

Unfortunately, word-level features are insufficient for complex problems. A , for instance, *denotes* a different concept than the literal denotation of a word (e.g., does Boston stand for the city of Boston or the Boston Red Sox?). The issue boils down to one of semantic tagging in the sentence [143].

Beyond document content, metadata can also be directly used as features, although the usage does tend to be domain-specific. For example, one could use email headers as good indicator of person names. Many news articles often start with a location name. Sometimes the purpose is to calibrate probabilities e.g., document URLs can be used to bias probabilities of entities. Again, the knowledge can be highly domain-specific. An interesting fact that has been noted is that names (e.g., bird names) have high probability to be a *project name* if the URL is from a computer science department domain than not.

Advances in Feature Engineering: Word Embeddings The preceding discussions highlighted the importance of feature engineering. In prior work, the utility of the features was central in determining the actual effectiveness of a machine learning algorithm for information extraction. In fairly recent work however, dating to the early part of this decade, word embeddings have emerged as an excellent way to mitigate feature engineering effort [118]. The basic idea behind most word embedding algorithms is to map each word in the corpus to a low-dimensional (in comparison to the dimensionalities of other text featurization algorithms like tf-idf), continuous (i.e., real-valued) *vectors*. The dimensions tend to lie between 20 and 100, depending on the size of the corpus. The embedding dimensionality is

---

[2]Aliases of an entity are the various ways the entity is written in a document e.g., Sir Elton John, E. John.

a hyperparameter and is accepted as given by the embedding algorithm. Generally, these embeddings work in a completely unsupervised fashion i.e. they do not require labeled data, although various complicated variants are also able to take labels into account when generating the embeddings. In modern work, a neural network (which is usually not deep unlike, say, a convolutional neural network) like a *skip-gram* model or a *continuous bag of words* model is used for the actual optimization.

The intuition is as follows. Imagine that we are *globally* trying to derive vectors for all the words such that, for a given word (e.g., 'cat'), either its vector is predictive of the vectors of its context words (in this case, the words that fall within a short distance of cat), or the context word vectors are together predictive of the given word.[3] However, when optimizing, the 'window' that defines the context is slid over all words in the corpus, so even though the context considered is local, the smooth sliding of the window ensures that influence is gradually percolating, leading to vectors that are hard to interpret but that have some remarkable intuitive properties.

These intuitive properties were borne out in some of the early (in the modern era[4]) work on word embeddings. The best known of these, and still extremely popular, is the *word2vec* algorithm. Word2vec allows optimization using either skip-gram or CBOW. Empirically, skip-gram has been found to be slightly better in actual problem settings. When using word2vec to embed a 'common' corpus like Wikipedia or the Google news corpus, it was observed that the vectors for words falling in the same (or similar) semantic class e.g., dog, cat and horse, fell close together in the vector space. Even more interestingly, it was possible to compute analogies in the vector space. In a famous example, one could do a computation such as **King − Man + Woman** in the vector space, and if a sufficiently large and broad corpus (like Wikipedia) was available then the resultant vector for the expression above was found to be close to the vector for **Queen**. This result is obtained despite the fact that the algorithm never received any labels, or any knowledge about which words fall in the same semantic class etc. Other algorithms inspired by word2vec or obeying similar methodology include GloVe and bag-of-tricks [84, 140]. All of them are based on some notion of context based on word co-occurrence, although the specific optimization functions are different. The bag-of-tricks approach also uses *sub-word* information, which allows it to deal with misspellings and OOV (out of vocabulary) words. This can be useful, especially when robustness is an important issue.

Embeddings have become so popular that they have percolated to allied communities in the knowledge discovery community, including graph embeddings, network embeddings, document embeddings and even knowledge graph embeddings. We cover the latter in Chap. 4. Because of the popularity of the field (formally called

---

[3] What is used for the prediction and what is being predicted is the difference between CBOW optimization and skip-gram optimization.

[4] The concept of embedding things into low-dimensional vector spaces is not novel. Using neural networks like skip-gram and CBOW and showing that they uncover properties of words that we intuitively understand is a relatively novel phenomenon.

'representation learning'), it has become an enormously influential and impactful research of area within machine learning. Most likely, we have not seen the last of it and work will continue to emerge in this area.

2.3.2 Relation Extraction

(RE) is the problem of detecting and classifying *relationships* between named entities (NEs) extracted from the text [9]. A relation usually denotes a well-defined (having a specific meaning) relationship between two or more NEs. We illustrate using the following examples (including the relation type with the arguments in parantheses):

1. Personal/Social: [Mary, Queen of Scots] was the royal cousin of [Elizabeth I]
2. Employment/Affiliation: [Albert Einstein] was one of the most distinguished faculty appointed for life at the [Institute for Advanced Study] in [Princeton, New Jersey].
3. Physical[5]: [Josephine] moved the [couch] from the left corner of the room to the right corner, so that it was next to her [aunt's portrait].
4. Geographical: [India] and [Nepal] are neighboring countries.

As with NER, we note that the set of relationship types that are within scope for the RE system is specified by a pre-defined ontology, although much of the existing work on relation extraction has tended to rely on a few ontologies for evaluation, such as the highly influential ontology from the Automatic Content Extraction (ACE) program [50]. ACE focuses on binary (relations between two entities), rather than arbitrary n-ary, relations. The two entities involved are generally referred to as *arguments*. The ACE ontology defines a set of major relation types and their sub-types, examples of major types including physical (e.g. an entity is physically near another entity), personal/social (e.g. a person is a family member of another person), and employment/affiliation (e.g. a person is employed by an organization) types. ACE also makes a distinction between relation extraction and relation *mention* extraction. The former refers to identifying the semantic relation between a pair of entities based on all the evidence we can gather from the corpus, whereas the latter refers to identifying individual mentions of entity relations. This is an interesting deviation from how NER systems have evolved, which (even today) focus primarily on the extraction of mentions, leaving it to modules like (both within-document and cross-document) coreference resolution and entity resolution (covered in Chap. 3) to cluster mentions into the same underlying entity. In practice, however, because

---

[5]This particular example shows how a seemingly simple sentence in English can prove enormously difficult to capture in a structured semantic form. *Movement* in this context is a quarternary relation (*who* moved *what* from *where* to *where*), but the sentence also expresses a simpler 'next-to' relation.

corpus-level relation extraction largely relies on accurate mention-level relation extraction, the latter is of primary interest in any discussion on relation extraction.

The examples also illustrate how open-ended the RE problem can be in terms of arguments, relation types and granularities. Even in the simple *geographical* example sentence, one can see that multiple sub-types of relations can exist (neighbor-of, has-continent, located-in-country etc.) and each such sub-type can be expressed in different ways (we could have equivalently written the sentence as 'India and Nepal share borders') rendering the ambiguity problem in IE particularly challenging for even supervised systems.

For these reasons, despite its obvious importance to domain-specific knowledge graph construction, relation extraction systems do not currently enjoy the same levels of performance as state-of-the-art NER systems [98]. Generally, the more complex the definition of an extraction, the worse its corresponding extractor performs. For example, event extraction, which we cover subsequently, performs even worse than relation extraction, with even state-of-the-art systems far from the 70% F-Measure mark (with even worse performance in more novel domains than the ones where the system might have been trained) that is considered feasible for large-scale adoption. Much research is still left to be done for these extraction problem domains.

In fact, successful RE requires detecting the *argument* mentions (e.g., Barack Obama, United States), with the entity types chaining these mentions to the ontological types (e.g., Politician, Country) their respective entities, and the type of relation (e.g., PresidentOf) that holds between these arguments. Relation extraction faces several challenges, several of which are shared by NER, though not to the same extent. First, RE is much more dependent on the domain, and the language, then NER. Supervised machine learning techniques applied to RE face the usual difficulty of a lack of sufficient training data. Another interesting problem is that the notion of a relation is inherently ambiguous, which means that labeling itself can be a problematic endeavor, reflected in high inter-annotator disagreements. Extending binary RE techniques to RE involving higher arity is also problematic, as we describe in the section on Event Extraction (EE). Even in the simple example above, we can intuitively see that detecting or classifying Barack Obama as President of the United States is much easier than detecting that Barack Obama was the President of the United States from 2009 to 2016. Quickly, the problem can become intractable, or the noise makes results unusable.

Many of the earlier techniques that we mentioned for NER, including supervised, semi-supervised and unsupervised learning, also apply to RE. Examples of systems that are now considered relatively classic include DIPRE, Snowball, KnowItAll and TextRunner [2, 32, 59, 187], all of which are semi-supervised rather than fully supervised or unsupervised. The problem is modeled differently however. Assuming binary relations, one way to model RE is to train either a binary (per relation) or multi-class classifier for pairs of extracted NEs. However, if one does this for every pair of entities extracted from the document, the complexity quickly becomes quadratic, and performance declines sharply. Thus, heuristics often have to be used to impose constraints. For example, one might only consider applying such classifiers to pairs of NEs extracted within a given span of text, or even within the same sentence. Recently, deep learning methods have also been extensively applied to RE [98].

Similar to NER, feature engineering has also been an important issue, and word-level, semantic and kernel features have all been proposed over the years. Kernel methods were especially popular in the earlier part of the last decade, an influential work being [189]. Feature classes specific to relation extraction have also been proposed. In general, such classes become necessary for 'higher-order' extractions like relations and events. Also, as described earlier, word embeddings have had a major impact on all of NLP, and RE feature engineering is no different. A more exciting development has been the *joint modeling and extraction* of relations and entities [190], and more generally, the joint extraction of events and entities. We describe the intuition behind such joint models in more details in the next section. Also, as pointed out earlier, deep learning methods, which almost always include some form of representation learning, have also been applied to RE, making the feature engineering problem less of an ad-hoc effort [98].

Because of the difficulty of the RE problem, and its recency, performance is lower across the board compared to NER; furthermore, performance tends to decline much more sharply as supervision levels are lowered compared to NER. More details on RE approaches and evaluation, including fairly comprehensive surveys and methodology reviews, may be found in [19, 98] and [9]. As the authors in that paper describe, the field is still relatively new compared to NER and much work is left to be done, especially in multi-lingual RE, n-ary RE and improvements in state-of-the-art methods.

2.3.3   Event Extraction

(EE) refers to the task of identifying events in free text and deriving detailed and structured information about them, ideally identifying who did what to whom, when, where, through what methods (instruments), and why [142]. Event extraction involves extraction of several entities and relationships between those entities. For instance, in an example taken from [142], extracting terrorist attack events from the text fragment 'Masked gunmen armed with assault rifles and grenades attacked a wedding party in mainly Kurdish southeast Turkey, killing at least 44 people.' involves identification of perpetrators (masked gunmen), victims (people), number of killed/injured (at least 44), weapons and means used (rifles and grenades), and location (southeast Turkey). Just like relation extraction, the problem can be very domain-specific. For example, in the mergers and acquisition domain, an event might be a merger that has just happened, which would require extracting the companies undergoing the merger (usually asymmetrically), the underwriter, the dates, specific merger terms, attorneys involved etc. EE is considered to be the hardest of RE, NER and EE.

For geopolitical style events, EE has been mainly studied using the ACE ontology [50], though other alternatives, such as CAMEO and ICEWS also exist [180]. For the biomedical domain, the BioNLP shared tasks are popular [91]. Because intense research continues to be conducted in EE, it is much too early to say which techniques are established and will stand the test of time. Some strands have started to emerge, however. For example, to reduce task complexity, early work tended to employ a sequence of classifiers that first extracted event *triggers*, then determined the trigger arguments [3, 28]. With the advent of deep neural networks, Convolutional Neural Networks (CNNs) have been employed as the pipeline classifiers [130]. Regardless, pipeline approaches suffer from error propagation via cascading, and *joint extraction* methods have emerged as state-of-the-art as a result [102]. As the name suggests, joint IE approaches tend to extract event triggers and arguments *together*, using methods such as structured perceptron [102], and dependency parsing algorithms [114].

The intuition behind joint IE is worth considering, especially considering the systems-level nature of knowledge graph construction. One can think about it as follows. Events and entities are *closely related*; entities are often actors or participants in events and events without entities are uncommon. The interpretation of events and entities is highly contextually dependent. Existing work in information extraction typically models events separately from entities, and performs inference at the sentence level, ignoring the rest of the document. An alternate approach that has recently come into vogue is to model the dependencies among variables of events, entities, and their relations, and to perform joint inference of these variables across a document. In essence, the learning problem is decomposed into three tractable subproblems: learning within-event structures, learning event-event relations, and learning for entity extraction. Probabilistic models are learned for all of these subproblems, with a joint inference framework integrating the learned models into a single model to jointly extract events and entities across a document.

The experimental results have been quite impressive, achieving state-of-the-art performance on EE typed according to the ACE ontology. Even more importantly, it was found that the benefits would often be mutual i.e. even a well-studied task like entity extraction benefited from better performance when done in the context of a joint inference framework. This is in line with the intuition stated earlier that there are close semantic connections between event triggers and arguments, and it is best to model such connections explicitly in the extraction framework itself.

One limitation of joint IE approaches is that they can suffers from complexity issues, like joint or collective approaches in general. To combat this problem, some approaches tend to rely on heuristic search to aggressively shrink the search space. One sophisticated exception is [154], which uses dual decomposition for joint inference with runtime guarantees. Other approaches proposing to do joint IE without enormous complexity burdens are continuing to slowly emerge from the research community.

Overall, the performance of EE is quite poor compared to NER. One avenue of research for improving EE performance is to exploit *document-level contexts*. Berant et al. [16] exploits event-event relations, e.g., causality, inhibition, which frequently occur in biological texts. It is clear that such relations are domain-specific. For more general texts, existing work tends to focus on exploiting temporal event relations [35, 54, 113]. For the ACE domain [50], there is work on utilizing event type co-occurrence patterns to propagation event classification decisions [81, 104]. Intuitively, co-occurrence is generally a useful feature (e.g., a DIE event tends to co-occur with ATTACK events but interestingly also, TRANSPORT events). In recent years, more general approaches have been proposed in this vein that can handle broad-domain event relations (e.g., causal and temporal) through the design of appropriate features. Similar to other extraction problems, ontological constraints can also be leveraged. For example, an entity mention of type PER can only fill roles that can be played by a person. The empirical utility of ontological constraints for such tasks is only starting to be studied in detail, and we will likely learn a lot more about the benefits and limitations of such constraints in the periods to come.

2.3.4   Web IE

The Web has emerged as the single biggest source of data for many domains, including reviews, e-commerce, academic literature and even investigative domains like securities fraud and human trafficking. Given a domain, one problem is to find and crawl relevant pages from the Web. This problem is known as domain discovery, and although recent research on it has made much progress (using techniques like reinforcement learning and page classification, for example), it continues to be an interesting and difficult area of study.[6] However, even given such a corpus, constructing a domain-specific KG involves extracting important pieces of information from the webpages.

Web IE [36], which covers this problem broadly, has an extensive history, with the vast majority of influential papers being published about 10–15 years within the initial growth of the Web. The dominant technique is a 'wrapper' [99], which was originally defined as a component in a Web information integration system aimed at providing a single uniform query interface to access multiple information sources. In the case where the information source is a Web server, a wrapper must query the Web server to collect the resulting pages via HTTP protocols, perform information extraction to extract the contents in the HTML documents, and finally integrate with other data sources. Due to historical reasons, the term 'wrapper' is now almost exclusively associated (in the IE community) with Web IE.

---

[6]Two reasons are the dynamic nature of the Web, but also the safeguards often put in place (such as registration requirements, and captchas) to avoid crawlers.

At a high level, wrapper induction (WI) is the process used to generate wrappers, usually using semi-automatic, rather than manual or fully automatic, methods. Broadly speaking, a wrapper performs a pattern matching procedure relying on a set of extraction rules. Tailoring a WI system to a new task can be challenging, depending on the text type, domain, and scenario. To maximize reusability and minimize maintenance cost, designing a trainable WI system has been an important topic in Web knowledge discovery and domain-specific search. Unlike NLP-centric IE, covered before, Web IE processes online documents that are semi-structured and is consumed by a server-side application program that is attempting to ingest the information in the documents into some kind of a database that is ultimately accessed as a knowledge graph. Unlike traditional IE, Web IE is often not able to leverage techniques such as lexicons and grammars to the same extent, since HTML is very different from natural language, and has to instead rely often on exploiting a mix of heterogeneous features, such as syntactic patterns, layout structures of template-based documents as well as more traditional text-based features.

Similar to NER systems, WI systems can be classed as supervised, semi-supervised and unsupervised [36], although sometimes the distinctions aren't completely clear.[7] Supervised WI systems take a set of web pages labeled with examples of the data to be extracted and output a wrapper. The user provides an initial set of labeled examples and the system (perhaps with the help of a GUI) may suggest additional pages for the user to label. The advantage of using a GUI is to empower general users, rather than programmers, to use the system and label additional data, which permits greater applicability. In this sense, supervised WI systems are different from supervised NER systems.

Regardless, labeling examples with precision has always been considered to be a difficult and arduous task in the broader AI community. Semi-supervised WI systems like OLERA and Thresher try to find a way around this problem by accepting a rough set of (instead of a complete and exact set of) examples from users for extraction rule generation [37, 77]. Systems like IEPAD do not require labeling [38], but instead push effort to the post-processing stage, when the user is asked to choose a target pattern and indicate the data to be extracted. All these systems are targeted for record-level extraction tasks. In the KG context, a record can be thought of as an entity of interest, along with attributes describing that entity. For example, a record describing a product would have the product identified as the central entity, and attributes like price or description would be attributes that constitute the non-ID 'columns' of the record. In this way, a KG can be incrementally constructed by collecting such record-level extractions over a corpus of webpages. However, since no extraction targets are originally specified for such systems, a GUI is still required for users to specify intuitive extraction targets after the learning phase.

---

[7]For example, if the system required a lot of model engineering but no training data, is it really 'unsupervised'?

Purely unsupervised Web IE systems do not use any labeled training examples and have no user interactions to generate a wrapper. The best examples are Road-Runner and ExAlg [6, 49], which were designed to solve page-level extraction tasks, while systems like DeLa and DEPTA are better-known for record-level extraction tasks. Unsupervised systems 'discover' the extraction target by segmenting the data that is used to generate the page or isolating non-tag texts in 'data-rich' regions of the page. The overall problem tends to be severely under-constrained, making it difficult e.g., several schemas may comply with the training pages leading to ambiguity as to which attributes are important and which ones are not pertinent to the domain. The choice of determining the right schema is left to users, meaning the system is not fully unsupervised after all. Similarly, if only some of the extractions are relevant, post-processing may be required for the user to select relevant data and name the extracted clusters appropriately. The general goal however is to ensure that supervision is minimal compared to semi-supervised and supervised WI systems. A downside of this automation is that the system may become overly dependent on the layouts of pages in the development and model engineering phases, and may not do well (or even crash) in test phases or when subjected to new domains. In other words, the price of high automation can sometimes end up being a lack of robustness and generalization.

Because unsupervised WI continues to be a difficult area of research, relatively few systems exist compared to supervised systems. Earlier, we described *Road-Runner* as a highly influential example, whose impact has continued to be felt many years after it was first proposed. RoadRunner considers the *site generation* process as encoding the original database content into strings of HTML code [49]. As a consequence, data extraction is considered as a *decoding* process. Therefore, generating a wrapper for a set of HTML pages boils down to the inference of a *grammar* for the HTML code. RoadRunner uses a matching technique to compare HTML pages of the same class and generates a wrapper based on their similarities and differences. The main idea is to start by comparing two pages, using the ACME (which stands for Align, Collapse under Mismatch, and Extract) technique, described in the original paper [49], to align the matched tokens and collapse for mismatched tokens. Since there can be several alignments, RoadRunner adopts *union-free regular expressions* to reduce the complexity of the process. The alignment result of the first two pages is compared to the third page in the page class, and the process continues.

Although the techniques details of RoadRunner can become complex, it is worth noting that, unlike other wrapper induction techniques that generated wrappers by examining labeled examples and has knowledge of the target schema, RoadRunner does not have prior knowledge about the organization of the pages. The technique is also quite efficient, since the authors proposed various mechanisms (such as union-free regular expressions) to ensure that the complexity does not exceed practical limits. In the original paper, RoadRunner was able to outperform supervised wrapper induction techniques like Wien and Stalker on an efficiency metric (CPU time) by orders of magnitude [49].

2.4 Evaluating IE Performance

Like any class of algorithms that has been rigorously researched and improved over decades, IE can also be evaluated on multiple metrics of interest (of its output) to a consumer. Many of these metrics quantify the difficulty of the problem via tradeoffs. Two metrics that we pay special attention to, and that will also play a role when we describe the evaluation of Entity Resolution in the next chapter are *recall* and *precision*, which were originally adopted from the Information Retrieval research community. These metrics, defined subsequently, can be respectively seen as measures of *completeness* and *correctness*.

Since the metrics will be computed on the output of an IE *with respect to* a gold-standard set of annotations, it is worthwhile asking what the outputs and gold-standard look like. Given a corpus of documents, each document d can be defined as a sequence of characters such that any (consecutive) *span* within the document may be identified using a *start offset* and an *end offset*. In this simple model, the gold-standard may be thought of as a set of triples, where each triple is of the form *(d, start offset, end offset)*. The IE output will also have this format. Let us define each element of the gold standard as a *slot*. Intuitively, the IE system ingests a corpus of documents, and outputs a set of candidate fillers for the slots. The goal is to evaluate the IE system's candidate slots against the reference slots in the gold standard. Although the NER systems that were covered fall very naturally in this category (with entity extractions filling slots), relation extractions and even event extractions (broken down into argument and trigger detection) can be defined in a similar way.

As a first step, let *numSlots* denote the number of slots in the gold standard G (a set of slots), and let the IE output O be the set of candidate slot fillers. We define the set of *true positives* to be $O \cap G$ i.e. the slots in O that were correctly extracted. Let the set of *false negatives* be $G - (O \cap G)$ i.e. the slots in G that were never extracted by the IE system. Finally, we refer to the slots in O that do not occur in G as *false positives*.

Using these definitions, precision and recall are defined as follows:

$$precision = \frac{true\ positives}{true\ positives + false\ positives} \tag{2.1}$$

$$recall = \frac{true\ positives}{numSlots} \tag{2.2}$$

This simple model of IE outputs and gold standard is appealing because it can easily be extended to other, more complex, IE problems. For example, in a particular kind of IE system (NER), the system must not only identify the slots but must also *type* the slots according to concepts from an ontology. Examples of concepts, as covered earlier, include ontological elements such as Person, Organization and Location. Per this notion, a slot in a gold standard would have the form *(d, start offset, end offset, type)*. For a more fine-grained picture of an IE system's performance, precision and recall are often measured for each slot type separately. The F-measure is used as a weighted harmonic mean of precision and recall, which is defined as follows:

$$F = \frac{2 * precision * recall}{precision + recall} \qquad (2.3)$$

Precision, recall and F-measure are metrics that are used quite broadly within computer science and are largely influenced by developments in the Information Retrieval (IR) community. However, some metrics are IE-specific. One example is the slot error rate (SER) which characterizes the extent to which the IE system makes mistakes (as opposed to the other metrics defined above, which characterize either correctness or completeness of the IE system). To define the SER, we must first assume that slots in the IE output O are *aligned* (usually algorithmically) to the slots in the gold standard G. A simple, but very conservative, definition of alignment is that both the start and end offsets must be equal for two slots to be aligned. Note that if the IE problem involves entity types, such as in NER, two slots (one from O and one from G) can be aligned but not match, since the type in each slot could be different. For example, it is possible that an extraction 'United States' got typed as a location by the IE system, but is actually a geopolitical entity in the gold standard. We assume that a slot in the gold standard can be aligned with at most one slot in G. With this in mind, let us define the variable *#wrong* as the number of slots in G that are (1) aligned with some slot in O, (2) do not match the aligned slot in O. In a similar vein, let us define *#missing* as slots in G that are not aligned with any slot in O. With these variables in place, SER can be defined as:

$$SER = \frac{\#wrong + \#missings}{numSlots} \qquad (2.4)$$

The intuition behind SER is relatively simple: how many of the slots in the gold standard G did the IE system either miss or get wrong? Similar to the other metrics, SER is always between 0.0 and 1.0, but unlike the other scores, a lower SER indicates a higher quality IE system.

Metrics other than SER, precision, recall and F-measure also exist for characterizing IE system quality. Most modern systems, however, tend to focus on these metrics.

2.5 Summary

Information Extraction (IE) is the first, and possibly most important step, in a domain-specific knowledge graph construction system, once preliminary steps such as domain discovery and dataset collection have been performed. Like most AI problems, IE is not a solved problem, though performance has continued to steadily improve over the years, including for semi-supervised and unsupervised IE. Modern advances, especially in event extraction, have been quite exciting, particularly due to deep neural networks, and more recently, generative adversarial networks (GANs). In the next several chapters, we cover downstream steps such as Entity Resolution and Knowledge Graph Completion that must consume the noisy outputs of IE systems.

Chapter 3
Entity Resolution

3.1 Introduction

Entity Resolution (ER) is the problem of devising *algorithmic* solutions for determining when two entities refer to the same underlying entity [66]. The problem is very common in almost all communities that deal with a lot of data, including knowledge discovery and data mining, databases and the Semantic Web. It is also a hard problem, despite (or perhaps because) of its *common-sense* nature, since it generally does not take specialized knowledge for a human being to answer the question of when two things are the same. ER problems widely exist in both industrial and non-industrial applications, and big technology companies often task entire teams to address the problem in its various guises. Multiple commercial and research solutions exist, some based on work that was originally done many decades ago [41]. Many books and special issues have also been dedicated to the topic. Although not humanly possible to cover ER in all its depth in this chapter, we attempt to synthesize the field in a conceptually meaningful way that will provide practical insights into why ER should be given special attention in any robust domain-specific KGC pipeline.

By way of a running example, consider the illustration in Fig. 3.1. Let us optimistically assume that complete and correct named entity recognition and relation extraction systems were applied to a corpus, yielding knowledge graph fragments. Clearly, the two nodes Nadal and Rafael Nadal extracted from the two documents need to be *resolved* since they are referring to the same underlying entity. In general, the problem is not unique to natural language sources, and can emerge even when we are constructing knowledge graphs over semi-structured (or even structured) raw sources like log files and XML. That being said, the natural language version of the problem is still special since one could potentially use linguistic clues to determine when extracted pronouns in a document refer to the same entity (anaphora or co-reference resolution). When extractions must be linked across documents, the problem is generally referred to as *cross-document coreference*

M. Kejriwal, *Domain-Specific Knowledge Graph Construction*, SpringerBriefs in Computer Science, https://doi.org/10.1007/978-3-030-12375-8_3

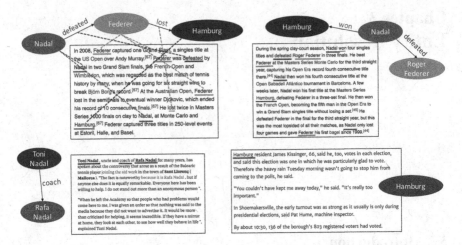

Fig. 3.1 An illustration of the cross-document ER problem. Only some extractions are shown from each of the documents, with the same color used if the nodes should be resolved together. The first document illustrates the 'provenance' of the KG nodes and relations. In general, each KG node or relation can trace its provenance to a *set* of *co-referenced* extracted mentions

resolution. Just because the document sources are separate doesn't preclude the use of linguistic cues and features.

However, the problem of ER is much broader, and the earliest known instances of it emerged in the patient linking and biomedical literature almost 50 years ago [123]. Even today, much more work has been done on the structured version of the problem (in both the database and Semantic Web communities) than in the natural language community [57, 96]. Much of the work in the structured data communities can be synthesized in a somewhat unified manner using similar concepts and terms. This synthesis, though brief, will be the focus of this chapter. Because of the nature of English language data as opposed to structured data, the NLP community has been forced to take a somewhat different approach to the problem. A good review of co-reference resolution may be found in [122].

In much work on ER, it was often the case that single-source, single-schema ER (often called *deduplication*) was the main focus of the research, along with close variants such as multi-source, single-schema ER. Recent research has attempted to address the multi-schema version of the problem [89], especially in KG-centric communities like the Semantic Web.

3.2 Challenges and Requirements

Before diving into solutions, we provide some intuition on why ER has proven so difficult to automate and algorithmically encode. Figure 3.2 provides some insight, despite its simplicity. The most important challenge in automating ER is the

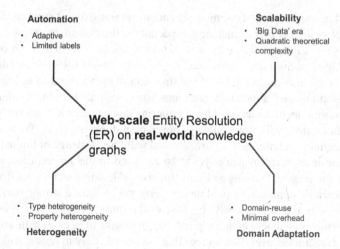

Fig. 3.2 An illustration of challenges that must be fulfilled to resolve entities in real-world knowledge graphs. Note that, even when graphs are not 'Web scale' a scalability challenge still arises because of the quadratic theoretical complexity of ER

ambiguity of the information extracted. Ambiguity is much harder to resolve in the presence of noise, and without access to underlying text, but even after accounting for those, it is not completely obvious how a machine is supposed to figure out that 'Nadal' in the first document refers to 'Rafael Nadal' or 'Toni Nadal'. In the much harder version of the problem, one would also have to figure out that 'Hamburg' in the first document is actually referring to the 'Masters Series Hamburg' and not the location Hamburg, as in the last document. More generally, because of nicknames (e.g., Pistol Pete vs. Pete Sampras) and other alternatives, one may also have a hard time coming up with viable candidates unlike the previous two cases, where one is faced with the finer-grained problem of choosing among viable linking candidates for a given node. Additionally, as in the natural language version of the problem, there is also the issue of having to deal with singleton nodes i.e. those that show up only once in the corpus and have no links to any other nodes.

Perhaps the most important challenge for an AI system attempting to counter the ambiguity in ER is that humans seem to draw on *background*, often intuitive, knowledge (often without even conscious reasoning) in several common ER domains that can be hard to pin down precisely in code. A computational challenge that will become more apparent in the subsequent discussion is *scale*, since naïve solutions to ER grow quadratically with the number of nodes in the KG [42]. As the introduction also pointed out, multi-schema ER is still very much in the nascent stages of research compared to the deduplication and single-schema cases. For knowledge graph construction systems that control the underlying reference ontologies of information extraction systems, multi-source, single-schema ER is still often applicable. However, in the most general case, multi-schema ER is necessary when constructing knowledge graphs over many sources, documents and tables, and across the outputs of multiple (not often transparent) systems. Building such an ER system is still a challenge, especially if the domain is unusual in some way and there is little guidance by way of prior work in academia, or precedent in industry. Good performance from machine learning-based ER systems (the state-of-the-art) may also require a lot of training data, which is hard to acquire, since regular sampling and annotation does not work well in ER due to *data skew* (intuitively, this can be understood by considering that one entity, if randomly paired with another entity, will almost always never be a duplicate pair). While modern machine learning techniques can *partially* deal with the challenge of limited labeled data, human-level performance is yet to be achieved in the general case, and data augmentation, transfer learning and semi-supervised learning techniques from other machine learning applications (and theory) have yet to make a strong mark on ER. Finally, noise in the input, usually because of the imperfections of IE systems but also due to incompleteness in the original data, also have to be dealt with, since KGs are rarely constructed over data sources that are already easy to reason with.

Given these challenges, it should not be surprising that real-world ER systems perform well in *some* aspects, such as automation or scalability, but may be deficient in others (e.g., heterogeneity) [85, 86]. Figure 3.2 captures these requirements visually.

Automation First, given the increasing expense of data scientists and subject matter experts, an ideal ER solution should exhibit a high degree of automation. This requirement *can* be met by a non-adaptive system, but such a system would have low robustness or real-world utility. If the system is adaptive and uses some form of machine learning, the requirement can only be fulfilled by algorithms that are minimally supervised (i.e. use small amounts of training data) or more rarely, completely unsupervised. An alternate option that has been explored in industrial ER is to leverage crowdsourcing or a professional annotation service. This option is limited by both cost and scalability.

Scalability The size and growth in data ecosystems like Wikipedia, social media, Linked Open Data, webpages, sensor data and schema.org markup (Chap. 5) suggests that building a feasible ER system requires devising solutions that meet requirements of elastic scalability, preferably requiring computational resources that increase only *linearly* in the size of the data. While for many algorithmic pipelines, this is an achievable goal, it is much harder for ER. The reason is that ER is inherently (and theoretically) *quadratic* as we subsequently describe. Bringing down this quadratic complexity to *almost-linear* complexity is a field of research in its own right (called blocking).

Heterogeneity Earlier, we already suggested that multi-schema ER is becoming more important for KG-centric applications. For the purposes of ER, multi-schema heterogeneity can be broken down into two separate (but inter-related) problems. The first is *type heterogeneity*, which arises when different ontologies are used for different raw data elements. For example, one IE system may produce fine-grained types such as *Inventor* and *Politician*, while another may produce coarser-grained types (e.g., *Employed Person* and *Unemployed Person*). The problem is further compounded by potential noise in type annotations, and by the presence of overlapping but not perfectly aligning type hierarchies across different sources and IE sub-systems. For example, is an inventor employed? The problem is more common than it seems, since the ontologies of many domains and datasets are developed relatively independently. Except in a few domains (such as the Gene Ontology in the biology domain [7]), heterogeneous ontologies and type-sets are the norm, rather than exceptions.

The second heterogeneity problem is *property heterogeneity* (the matching of property or edge labels across ontologies) that tends to arise once types are aligned. For example, let us assume that an ER system has correctly managed to address type heterogeneity by aligning *Inventor* (in one ontology) with *Entrepreneur* (in a second ontology). The ER system would also have to deduce such alignment relationships between properties such as *:co-founder_of* and *:organization*. As these examples show, alignment does not necessarily imply relationships of subsumption or equivalence, but is simply an empirical determination of sufficient entity overlap. Just like other processes in KGC, like IE, *instance-driven ontology alignment* is itself a problem that continues to be researched and has not been solved with human-level performance [1].

Domain-adaptation Finally, if the ER system is to be *re-used* across domains, it must also be *domain-adaptable* in its workings. By domain-adaptable, we do not mean that the ER system has to be a static one-size-fits-all model that magically works well across all domains, or even that it needs to be trained in one domain but is expected to perform well in a separate test domain (transfer learning). Rather, it must have the ability to adapt as the domain changes. In this sense, domain-adaptability is not necessarily mutually exclusive from domain-specificity, but refers to the *meta-ability* of an ER system to be re-trained, re-deployed and re-used on a different domain with *minimal overhead*. Domain-adaptability is hard to formalize; it is a practical and empirical constraints. In practice, no ER system is completely domain-adaptable (some assumptions built into the system are directly influenced by a use-case) or completely domain-specific (some re-use is always possible, and more re-use is generally possible in related domains). However, some systems are so strongly influenced by a particular use-case (e.g., product Entity Resolution) that adapting them to other domains is equivalent to writing the system from scratch. Event resolution is emerging as an excellent example of this phenomenon. Although event resolution is still heavily in flux as a research area, with a growing body of output, the best systems (both for event resolution and extraction e.g., BBN ACCENT [153]) tend to be heavily tuned not only for events, in general, but specific *types* of event. It is not unreasonable to suppose that an event resolution system designed for geopolitical events may not do as well if transferred to concert or entertainment events. Characterizing and evaluating such transferability is currently an open research problem.

In summary, there is a natural tradeoff between domain-adaptability and automation, and the two tend to influence each other in the design phase. For domain-specific ER systems, it is unlikely that (even with reasonable training data) the system will be able to resolve entities that look very different from the entities the system was designed to resolve. Modern *representation learning* techniques, such as word and graph embeddings [90, 132], have alleviated concerns about domain adaptation to a certain extent, since embeddings can be trained on unlabeled corpora. There is no free lunch however, since training good embeddings requires a sufficiently large corpus. In some domains, availability of such corpora may be limited.

3.3 Two-Step Framework

Even in early research, the quadratic complexity of pairwise ER was well recognized [123]. Given two data sources G_1 and G_2, where the set of non-literal entities in graph G is represented by the symbol E, a naïve ER system would evaluate all possible entity pairs. Assuming constant cost per evaluation, the run-time would be $O(|E_1||E_2|)$. In the rest of this section, for two entity sets E_1 and E_2, an entity pair (e_1, e_2) is denoted as bilateral iff $e_1 \in E_1$ and $e_2 \in E_2$. Given a collection of entities from $E_1 \cup E_2$, two entities e_1 and e_2 are said to be bilaterally paired iff (e_1, e_2) is bilateral.

To mitigate the quadratic complexity of generating all possible bilateral pairs, a two-step approach is adopted, as illustrated in Fig. 3.3 [41]. The first step, *blocking*, uses a many-many function called a blocking key to cluster approximately similar entities into overlapping blocks [42]. Only entities sharing a block are bilaterally paired and become candidates for further evaluation by a link specification function in the similarity step [172]. The link specification function may be either Boolean or probabilistic, and is used to indicate whether a candidate entity pair represents the same underlying entity.

Because ER developed as an important research area in the database community, the majority of ER research still assumes input databases to be structurally homogeneous i.e. if more than one database is input to the ER system, the databases

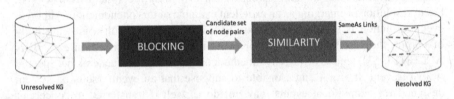

Fig. 3.3 The typical two-step workflow adopted for Entity Resolution

are assumed to have the same schema and same semantics [41,57]. In the knowledge graph world, this would be equivalent to matching entities between knowledge graphs that have the same underlying ontology i.e. sets of concepts and properties. An important special application of structural homogeneity is deduplication, whereby matching entities in a single dataset must be found. Although structural homogeneity may seem like a limitation (which in some applications, can be severe), it is also often the case that ER is the next step after information extraction in a domain-specific KGC pipeline, and a single ontology is involved. Thus, the goal of ER is to deduplicate sets of entities extracted and typed according to this ontology. In the rest of this section, structural homogeneity is assumed. Later, we will briefly discuss extending the two-step model to include structural heterogeneity, but for an extensive discussion refer the interested reader to [86].

3.3.1 Blocking

Blocking is a preprocessing step that is used to mitigate the quadratic complexity of applying the link specification function on all (unordered) pairs of mention nodes in the knowledge graph. Given a set M of mention nodes (i.e. 'raw' entities extracted from documents), this *exhaustive set* contains $(|M||M| - 1)/2$ distinct unordered pairs, which is an untenable number of link specification computations for $|M| \gg 1000$. In the most general case, blocking methods use a many-many function called a *blocking key* to cluster approximately similar entities into overlapping blocks.

Definition 3.1 (Blocking Key) Given a set M of mention nodes, a blocking key K is a many-many function that takes a mention $m \in M$ as input and returns a non-empty set of literals, referred to as the blocking key values (BKVs) of m.

Let $K(m)$ denote the set of BKVs assigned to the mention $m \in M$ by the blocking key K. Furthermore, without loss of generality, the literals in the definition above are all assumed to be strings.

Example 3.1 (Blocking Key) Assuming the publication domain, with the *Publication* concept being the domain of properties Author, Venue and Year, and with the special property : *label* indicating the title of the publication, one possible blocking key K for deduplicating citations might be $overlap(Author(m_1), Author(m_2)) \wedge commonToken(Venue(m_1), Venue(m_2))$. This rule says that two publication mentions should share a block if their titles have at least three common tokens, *or* their venues have a common token (e.g., ACM KDD vs. KDD). We return to this example later in the context of automatically 'learning' good blocking keys. Note that this blocking key can generate multiple blocking key values for each node. More precisely, if a mention node has j authors and t tokens in its venue, the number of possible BKVs for the node is $j + t$. If any of these $j + t$ BKVs intersect with the BKV set of another node, they would fall within the same block (labeled by its BKV). The two nodes would share more than one block if they share more than one BKV. Intuitively, this would happen when they share more than one common token across their venue attributes, or they share more than one author, or they share an author *and* a common token in their venue attributes (or any combination of the three options). For reasons covered shortly, some blocks may end up being discarded. In many situations, therefore, the higher the number of shared blocks between two mentions, the higher the probability they will actually be compared in the similarity step.

Given a blocking key K, a *candidate set* $C \subseteq M \times M$ of mention pairs can be generated by a *blocking method* using the BKVs of the mentions. We describe three influential methods that are generally included in established surveys [42], and all of which assume that a blocking key K is already specified by a user. Depending on the method, K must also obey some constraints. Subsequently, we also describe the automatic *learning* of good blocking keys.

3.3.1.1   Traditional Blocking

Given a blocking key K, an obvious solution is to generate the candidate set C as the set $\{(m_i, m_j)|m_i, m_j \in M \wedge m_i \neq m_j \wedge K(m_i) \cap K(m_j) \neq \{\}\}$. Put simply, if two mention nodes share a blocking key then they would be paired and inserted into C. Note that the definition of C as a set further implies that m_i and m_j may share multiple BKVs, although in the earliest definitions of this so-called *traditional blocking*, a mention was allowed to have exactly one BKV (hence, blocks could not overlap but represented a partition, with singleton blocks automatically discarded from further comparison).

A problem with traditional blocking approach is that of *data skew*. Consider, for example, two mentions from a *People* knowledge graph that are blocked based on the tokens in their last names. Last name frequencies in many countries tend to exhibit skew (a Zipf-like distribution) for some values (e.g. *Smith* in English-speaking countries). A consequence of the skew is that the run-time of the blocking method ends up being roughly proportional to the number of pairs generated by the largest block. This implies that run-time is still roughly quadratic in the number of mentions, unlike state-of-the-art blocking methods, where run-time tends to be slightly super-linear.

Despite this problem, traditional blocking is often the first line of attack in practical systems. In recent years, researchers have modified traditional blocking to handle the large blocks that result from skew. A simple method that is easy to implement and difficult to outperform is *block purging*. The premise of the method is that, with a sufficiently expressive blocking key, blocks that are too large can be safely ignored. Such blocks are most likely indexed by BKVs that are equivalent to stop-words like *the* or *an*. The algorithm takes a purging threshold as an input parameter, and discards all blocks that have more pairs than this threshold. The threshold may be learned from the data, and tends to be empirically robust to good default values as long as the default value is not too low.

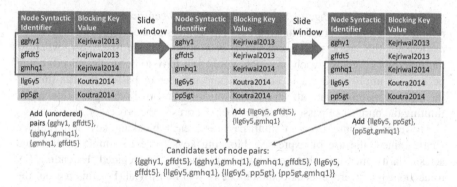

Fig. 3.4 An illustration of the Sorted Neighborhood workflow

3.3.1.2 Sorted Neighborhood

Another influential blocking method that was fundamentally designed to *guarantee* a bound on the size of the candidate set is the Sorted Neighborhood (SN) method, also known as *merge-purge* [76]. The algorithm works as follows. First, a single blocking key value (BKV) is generated for each mention using a many-*one* blocking key. Next, the BKVs are used as sorting keys to impose an ordering on the mentions. Finally, a window of constant size w is slid over the sorted list. All mentions sharing a window are paired and added to the candidate set. Figure 3.4 illustrates a workflow with a sliding window of size 3. We assume that the single BKV is generated by concatenating the last name of the first author of the publication with the year of the publication.

The sliding window has two implications for candidate set generation. First, mentions with *different* blocking key values may *still* get paired. This happens when the window straddles mention IDs in the list that have consecutively sorted BKVs (e.g., $gffdt5$ and $llg6y5$ get paired in Fig. 3.4). Second, some mentions with the *same* blocking key value may *not* get paired. For example, in Fig. 3.4, if the BKV for node $gmhq1$ had been Kejriwal2013 instead of Kejriwal2014, and the window size had been 2 instead of 3, then the node pair $\{gghy1, gmhq1\}$ would not be added to the candidate set.

Assuming that the window size w is much smaller than the total number of mentions, Sorted Neighborhood has time and space complexity that is *linear* in the size of the data. For this reason, it has endured as a popular blocking technique, especially when inputs are highly structured and it is possible to devise good blocking keys that yield a single, reliable BKV per mention. Numerous variations now exist, including implementations in Big Data architectures like Hadoop and MapReduce [95]. In general, the primary differences between the variants and the original version are input data types (e.g., XML Sorted Neighborhood vs. Relational), constraints on blocking keys and tuning mechanisms for the sliding window parameter (e.g. adaptive vs. constant) to achieve maximal performance in the similarity stage.

The main disadvantage of SN algorithms for KG-centric deduplication is their reliance on a single-valued blocking key. The authors of the original SN algorithm recognized this as a serious limitation and proposed *multi-pass SN*, whereby multiple blocking keys (each of which would still have to be single-valued) could be used to improve coverage. For a constant number of passes, the run-time of the original method is not affected asymptotically. Practical scaling is achieved by limiting the number of passes to the number of cores in the processor.

However, because even in multi-pass SN, each blocking key still remains single-valued, the use of expressive blocking keys (or even simple token-based set similarity measures that have high redundancy) is precluded. Extending SN to account for heterogeneous data sources is also non-trivial. For this reason, the application of Sorted Neighborhood to knowledge graphs and other heterogeneous, semi-structured data sources has been limited. The use of a simple blocking method such as traditional blocking (combined with skew-compensating measures like block purging) has remained popular for that reason.

3.3.1.3 Canopies

Clustering methods such as *Canopies* have also been successfully applied to blocking [112]. The basic algorithm takes a *distance function* and two threshold parameters $tight \geq 0$ and $loose \geq tight$, and operates in the following way. First, a seed mention m is randomly chosen from M. All mentions that have distance less than $loose$ are assigned to the *canopy* represented by m. Among these mentions, the mentions with distance less than $tight$ (from the seed mention) are removed from M and not considered further. Another seed mention is now chosen from all mentions still in M, and the process continues till all points have been assigned to at least one canopy.

In the Canopies framework, each canopy represents a block. However, unlike more typical methods like Sorted Neighborhood, Canopies does not rely on a blocking key, and instead takes a distance function as input. For this reason, at least one work has referred to it as an 'instance-based' blocking method, and distinguished it from 'feature-based' blocking methods such as Sorted Neighborhood and Traditional Blocking.

Similar to other popular blocking methods like Traditional Blocking and Sorted Neighborhood, several variants of Canopies have been proposed over the years, but the basic framework continues to be popular. For example, a nearest-neighbors method could be used for clustering mentions, rather than a threshold-based method. In yet another variant, a blocking key can be used to *first* generate a set of BKVs for each mention, and Canopies can then be executed by performing distance computations on the *BKV sets* of mentions, rather than directly on the mentions themselves. Because this variant relies on a blocking key, it can no longer be considered an instance-based blocking method.

For the distance function, the method has been found to work well with (the distance version of) a number of token-based set similarity measures, including Jaccard and cosine similarity [12], but in principle, many other distance functions can be used.

3.3.1.4 Research Frontier: Learning Blocking Keys

Earlier, we presented an example of a blocking key $overlap(Author(m_1),$ $Author(m_2)) \wedge commonToken(Venue(m_1), Venue(m_2))$. This key, while intuitive, has some severe drawbacks. First, it would cluster together all papers authored by the same author into one single block. Some authors have many hundreds of papers, and some others collaborate with others who also have hundreds of papers. It is quite likely that, by itself, a rule such as this would end up placing a large number of publication mentions in a single block, which would negate the complexity benefits of blocking. A similar problem occurs with the second part of the rule, which says that a common token in venues is all that's required for two publication mentions to get blocked together. Tokens like ACM and IEEE are very common in venue titles (at least in Computer Science and Engineering), and once again, we would face the problem of having far too many mentions (the vast majority of which are non-matching) placed in one block.

In general, we note that the problem of *data skew* cannot really be avoided, unless the rules are extremely precise. Because there is a tradeoff between precision and recall in most real-world AI systems, the complexity reductions entailed by blocking would end up having a high cost in terms of lost recall. The goal of blocking always is to try and reduce complexity with minimal loss in recall. The blocking methods do provide some respite from data skew, if tuned correctly. For example, block purging would remove blocks that have too many mentions from further consideration in the similarity step. However, the problem with the blocking rule above is that a 'big' block would also contain many matching pairs along with non-matching pairs of mentions. Removing a big block would reduce complexity, but would yield recall that is almost trivially low.

This argument shows that, even with good blocking methods in place, the *quality* of the blocking key itself is very important for achieving a good tradeoff between recall and complexity reduction. Devising such a blocking key was once the turf of domain experts and knowledge engineers (and in many domains, still is), but with the advent of machine learning, it has been found that good rules can be learned automatically using a training set of labeled duplicate and non-duplicate mention node pairs.

The general idea is to frame the problem as that of learning rules in *Disjunctive Normal Form* (DNF). DNF formulae can technically be used to represent any propositional formula, but in practice, some restrictions are imposed (e.g., negations of blocking predicates are not allowed, since this could make the blocking step intractable). The optimization function can be informally stated as that of learning a DNF formula such that: (1) the formula yields True for pairs of nodes in the positive training set, and (2) the formula yields False for pairs of nodes in the negative training set. It turns out that this problem can be decomposed as the famous *Red-Blue Set Covering* problem, which is known to be NP-hard. However, the problem is well-explored in the literature, and even a relatively greedy approach offers some good guarantees. In the few (but still growing) literature on blocking key learning, greedy algorithms were used to solve the reduced version of DNF blocking key learning. For more details, we refer the reader to fairly recent work in [87, 88] and [148].

3.3.2   Similarity

Once obtained, the candidate set C of mention pairs must undergo similarity computations to determine, whether probabilistically or deterministically, the subset of C that comprises duplicate mention pairs. In an i.i.d (independent and identically distributed) formalism, each mention pair can be independently assigned a score, with higher scores indicating greater likelihood of the pair being a duplicate pair. Although scores are typically normalized so that they lie between [0, 1], there is controversy about interpreting them as probabilities. We sidestep this controversy by continuing to refer to these numbers as scores.

Two issues now remain, one of which is concerned more with practice, and the other with theory. First, what methods do we use to obtain the scores in the first place? Intuitively, the 'goodness' of every such method should be measured by comparing against the perfect outcome i.e. the ground-truth. An ideal method that has knowledge of the complete ground truth would assign a score of 1.0 to every duplicate pair, and 0.0 to every non-duplicate pair. Subsequently, we present some formal methods for measuring performance using this principle. In practice, a validation or development set of labeled pairs can be used to select and tune methods to yield (by way of expectation) the best performing distribution of scores.

Second, given that scores output by a practical method will lie between 0.0 and 1.0, and will not necessarily be binary, how should one use these scores to 'partition' the set C into sets of duplicate pairs and non-duplicate pairs? There is a well-known theoretical model in the early ER literature known as the *Fellegi-Sunter model*, named after the scientists who first formulated it. The model generally requires not one but *two* (not necessarily distinct) thresholds to achieve a desired optimal tradeoff between the often conflicting goals of minimizing false positives and false negatives (which affect both *precision* and *recall*; see Sect. 3.4). The intuition behind using two thresholds is that they partition the set of mention pairs into three sets (matches, non-matches and *possible* matches i.e. pairs requiring manual review). To compute the score that will be compared to these thresholds, the ratio of conditional probabilities (with the condition based on whether the pair if assumed to be matching or non-matching) is used. For more details on the Fellegi-Sunter model, we refer the reader to [61].

We also note that, for models that rely on rule bases or heuristics, labels may be directly output. However, to get good rules or heuristics, extensive domain engineering effort is required and in recent years, such methods have been largely superseded by machine learning. Therefore, we focus on machine learning methods for assigning scores to pairs in the candidate set. The evolution of the ER field (not necessarily within knowledge graphs or Semantic Web alone) is complicated; we provided an extensive survey, and the limitations of existing work, in [86].

In machine learning-based ER, each mention pair in C is first converted to a numeric (typically, but not necessarily, real-valued) *feature vector*. Figure 3.5 illustrates the procedure for two mentions, assuming that special (i.e. 'dummy') values are used in the event that (1) values for a given property are missing from both mentions, even though values for that property were observed for at least one other mention in the dataset; (2) the value for a given property is missing from one (but not both) mentions.

In general, given n properties, and m functions in the feature library, the feature vector would have mn elements. We say general, because it is also possible that some features are designed for specialized values (e.g., a feature that computes the number of milliseconds between two date values), and not applicable to two

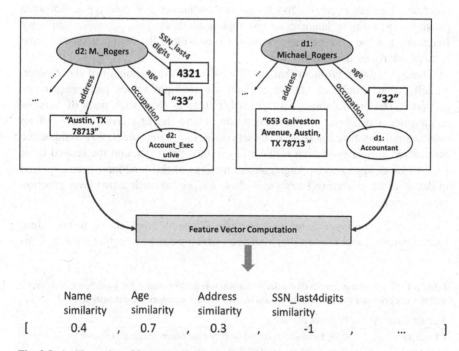

Fig. 3.5 An illustration of feature vector computation (between the two nodes mentioning Michael Rogers) assuming structural homogeneity. -1 is the dummy value used in this example

arbitrary values. However, having so many features, many of them correlated and often not useful, can be detrimental to machine learning generalization, especially when the training set is small and highly heterogeneous, as is the case with real-world ER tasks. There are several remedies for this; we consider two of the most popular ones. First, one can start by computing all possible (i.e. mn) features and then apply a feature selection method like Lasso. Second, one could spend some amount of domain engineering effort assigning only a few (in many cases, one might be sufficient) features to each property. Assuming that at most c feature function are assigned to a column, with $c \ll m$, the total length of the feature vector will be much less than mn, which would lead to presumably faster generalization and more robust performance. Often, domain expertise can be leveraged to limit m to 1 by deciding which feature function might be best for a given property. Considering again the example in Fig. 3.5, we may have decided to use NYSIIS for computing name similarity, normalized age difference for computing age similarity, TF-IDF for computing address similarity and so on, as opposed to applying every single function in our library for every possible property feature computation. Intuitively, one would not want to apply NYSIIS to address similarity since it would likely not be useful (and may even cause noise and problems with generalizing on fewer training examples, an important concern).

What feature functions should be included in a library? There is an enormous body of work on both string similarity, and to a lesser degree, phonetic similarity functions, and not nearly as much research on numeric or date types. Software packages in multiple languages exist that implement many of these similarity functions. For the sake of completeness, we provide a list of functions that have been popularly used in Table 3.1.

There is another problem that we alluded to at the beginning of the chapter, namely, what should we do when there are multiple values per mention per property? The pre-dominant way to extract features from such pairs of 'sets' is to consider a two-layer similarity function, where the first layer consists of an atomic similarity function (e.g., if the set consists of string values, this could be the normalized similarity version of the edit distance function), and the second layer consists of an aggregation. Aggregation functions of this kind have been explored in detail in the clustering literature. Below, we portray such a two-layer function using an example.

Example 3.2 (Two-layer Similarity Feature) Consider two sets of names {Jim, Jimmy, Jeremy} and {James, Jim} between which we need to output a single simi-

Table 3.1 Illustrative instances of similarity functions typically used in ER workflows. Neither the feature categories nor the example functions in each feature category are exhaustive

Feature category	Example functions
Character	Edit, Levenstein, Affine Gap, Smith Waterman, Jaro, Q-gram
Token	Monge Elkan, TF-IDF (Soft, Q-gram), Jaccard
Phonetic	Soundex, NYSIIS, ONCA, Metaphone, Double Metaphone

larity score. An atomic similarity feature could be Monge-Elkan. Namely, we could compute Monge-Elkan scores between each pair of terms in the two sets and output the results as a *complete weighted bipartite graph*. Several aggregation measures could be considered for the second layer of the similarity feature. For example, we could take either the minimum or maximum of all scores in the graph as the final score. We could also take the average. A more robust mechanism that has been found to work well in several cases is the Hungarian algorithm, which tries to match each term in the first set with at most one term in the second set (vice versa) such that the total sum of scores is maximized. We only keep those edges in the graph that were included in the optimal matching output by the Hungarian algorithm. Note that the number of edges in the graph will be the minimum of the cardinalities of the two sets, since a term in any set can never receive more than one assignment. In this sparser graph, containing only optimal assignments, we could take the average, minimum or maximum (or any reasonable aggregation) of edge weights.

As the example above shows, the more complicated a similarity measure becomes, the more degrees of freedom it tends to have, and the more options there are to explore. The extensive literature on ER is a good place to look for defaults, but for unusual domains, there is no substitute for careful tuning, some of which may have to be done through a systematic process of trial and error. Lately, vector space embeddings (covered in the next chapter) have alleviated some of the feature engineering effort that has gone into a typical ER workflow, but there is still much more work to do on this front.

Once a feature extraction methodology is in place, each mention pair in C is, in turn, converted to a feature vector. A machine learning classifier is trained on positively and negatively labeled training samples, and is thereby used to assign scores to the vectors in the candidate set. Several classifiers have been explored in the literature, with random forest, multilayer perceptron and Support Vector Machine (SVM) classifiers all found to perform reasonably well. We note that, although all of these classifiers make the i.i.d. (independent and identically distributed) assumption, transitivity does play a strong role in real-world ER determinations (if (e, j) and (e, k) are classified with high scores, it is reasonable to suppose that so should (j, k)). This fact is typically employed, not at this stage, but in the post-processing clustering and soft transitive closure stage (briefly discussed in a subsequent section) where we take the outputs of similarity and attempt to 'collapse' them into clusters, with each cluster representing all mentions of a single underlying entity.

3.4 Measuring Performance

The independence of blocking and similarity suggests that the performance of each can be controlled for the other in experiments. In the last decade, in particular, both blocking and similarity have become increasingly complex. It is the norm, rather than the exception, to publish either on blocking or on similarity in an individual publication. Despite its potential disadvantages (in practice, there are interdependencies between blocking and similarity, since feature functions and biases could often be traded between the two, sometimes without knowledge), this methodology has resulted in the adoption of well-defined evaluation metrics for both blocking and similarity. This independence assumption has been challenged in a small number of applications in recent years; as just one example, a blocking technique called *comparisons propagation* proposes using the outcomes in the similarity step to estimate the usefulness of a block in real time [137]. The premise is that if a block has produced too many non-duplicates, it is best to discard it rather than finish processing it. By this logic, the cost of processing the block outweighs the gain, at least in expectation.

While such techniques are appealing, their implementations have mostly been limited to serial architectures, owing to the need for continuous data-sharing between the similarity and block generating components. Experimentally, the benefits of such techniques over independent techniques like Sorted Neighborhood or traditional blocking (with skew-eliminating measures such as block purging) have not been established extensively enough to warrant widespread adoption. The two-step workflow, with both steps relatively independent, continues to be predominant in the vast majority of ER research. With this caveat in place, we describe these metrics below.

3.4.1 Measuring Blocking Performance

The primary goal of blocking is to scale the naïve one-step ER that pairs all mentions (order-independently) with each other. A blocking system accomplishes this goal by generating a smaller candidate set. If complexity reduction were the *only* goal, the blocking system could simply generate the empty set and obtain optimal performance. Such a system would be useless because it would generate a candidate set with zero duplicates coverage.

Thus, duplicates coverage and candidate set reduction are the two goals that every blocking system seeks to optimize. To formalize these measures, let Ω be denoted as the *exhaustive set* of all $^{|M|}C_2$ pairs; in other words, the candidate set that would be obtained if there were no blocking. Let Ω_D denote the subset of that contains all (and only) matching mention pairs (i.e. semantic duplicates). Ω_D is designated as the ground-truth or gold standard set. As in previous sections, let C denote the candidate set generated by blocking. Using this notation, *Reduction Ratio (RR)* is defined by the equation below:

$$RR = 1 - \frac{|C|}{|\Omega|} \tag{3.1}$$

The higher the Reduction Ratio, the higher the complexity reduction achieved by blocking, relative to the exhaustive set. Less commonly, RR can also be evaluated relative to the candidate set C_b of a *baseline* blocking method (by replacing Ω in Eq. 3.1 with C_b). Note that, since RR has quadratic dependence, even small differences in RR can have an enormous impact in terms of run-time. For example, if Ω contains 100 million pairs (not an unreasonable number, since it would only take a mentions set M with about 20,000 mentions i.e. a relatively small dataset), and System 1 achieves an RR of 99.7%, while System 2 achieves 99.5%, their candidate sets would differ by 200,000 pairs.

In a similar vein, coverage, or Pairs Completeness (PC), is defined below:

$$PC = \frac{|C \cap \Omega_D|}{|\Omega_D|} \qquad (3.2)$$

One interpretation of PC is to consider answering the following question: *if we knew $\mathscr{L}$ and apply it to the candidate set C output by blocking, what would be the recall?* From this perspective, PC is nothing but a measure of recall (used for evaluating overall duplicates coverage in the similarity step, as described in the subsequent section) that *controls* for the errors in further learning or approximating $\mathscr{L}$, which is *not* known. In other words, Pairs Completeness gives an *upper bound* on the recall metric. For example, if PC is only 80%, meaning that 20% of the duplicate pairs did not get included in the candidate set, then coverage on the full ER task will never exceed 80%.

There is typically a tradeoff between achieving high PC and RR. The tradeoff is achieved by tuning a relevant parameter. There are two ways to represent this tradeoff. The first is a single-point estimate of the *F-Measure*, or harmonic mean, between a given PC and RR:

$$F - Measure = \frac{2 \times PC \times RR}{PC + RR} \qquad (3.3)$$

A single-point estimate is only useful when it is not feasible to run the blocking algorithm for multiple parameter values. Otherwise, a more visual representation of the tradeoff can be achieved by plotting a curve of PC vs. RR for different values of the parameters.

Another tradeoff metric, Pairs Quality (PQ), is less commonly used than the F-Measure of PC and RR:

$$PQ = \frac{|C \cap \Omega_D|}{|C|} \qquad (3.4)$$

Superficially, PQ seems to be a better measure of the tradeoff between PC and RR than the F-Measure estimate, which weighs RR and PC equally, despite the quadratic dependence of the former. In this vein, PQ has been described as a precision metric for blocking. Intuitively, a high PQ indicates that the generated blocks (and by virtue, the candidate set) are dense in duplicate pairs.

In practice, PQ gives estimates that are difficult to interpret, and can be misleading. For example, suppose there were 1000 duplicates in the ground-truth, and only contained 10 pairs, of which 8 represent duplicates. PQ, in this case, would be 80%. Assuming that the exhaustive set is large enough that RR is close to 100%, the F-Measure (as defined above) would still be less than 2% (since PC is less than 1%). The F-Measure result would be correctly interpreted as an indication that, for practical purposes, the blocking process has failed. The result indicated by PQ alone is clearly misleading, suggesting that, as a tradeoff measure, PQ should not be substituted for the F-Measure of PC and RR. An alternative, proposed by at least one author but (to the best of our knowledge) not used widely, is to compute and report the F-Measure of PQ and PC.

3.4.2 Measuring Similarity Performance

Given a candidate set C, the similarity step uses a learned linking function to partition C into sets C_D and C_{ND} of duplicates and non-duplicates respectively. The two metrics predominantly used for evaluating the similarity step, and by virtue, ER as a whole, are *precision* and *recall*:

$$Precision = \frac{|C_D \cap \Omega_D|}{|C_D|} \tag{3.5}$$

$$Recall = \frac{|C_D \cap \Omega_D|}{|\Omega_D|} \tag{3.6}$$

In other words, precision is the ratio of true positives to the sum of true positives and false positives, while recall is the ratio of true positives to all positives in the ground-truth. Similar to PC and RR defined earlier, there is a tradeoff between achieving high values for precision and recall. An F-Measure estimate can again be defined for a single-point estimate, but a better, more visual, interpretation is achieved by plotting a curve of precision vs. recall for multiple parameter values.

Note that, since similarity is defined as a binary classification problem in the machine learning interpretation of ER, other measures such as accuracy can also be defined. One reason why they are not considered in the ER literature is because they also evaluate performance on the negative (i.e. non-duplicates) class, which is not of interest in ER. An alternative to a precision-recall curve is Receiver Operating Characteristic (ROC), which plots true positives against false positives. Historically, and currently, precision-recall curves dominate ROC curves in the ER community, but nowadays, important machine learning packages (e.g., sklearn in Python) allow a user to print out various metrics and curves without any programming. In real life, we recommend printing out both the precision-recall and ROC curves to evaluate both (1) how well the ER system is doing in an 'absolute' sense; (2) how well the ER system is doing above *random*.

3.5   Extending the Two-Step Workflow: A Brief Note

Although the vast majority of ER (including research and implementation) is concerned with optimizing and automating one or both of the two steps in the standard blocking-similarity workflow, the heterogeneity of knowledge graphs can require two additional steps to be given some thought in some application domains. Earlier in the chapter, we discussed how heterogeneous schemas (by way of type and property heterogeneity) can cause problems for ER systems. If we are linking mentions between two independent knowledge graphs with different schemas, or even between mentions in a single knowledge graph with very fine-grained types and properties, it is important to develop a robust type and property matching system that can be executed prior to blocking to reconcile schema heterogeneity. Just like blocking, we generally desire such preprocessing steps to be recall-friendly, since we do not want to risk losing (already sparse) duplicates before the similarity step, which is expected to add noise of its own.

It turns out that there is a large body of work on both type and property matching, sometimes involving the same researchers as ER, and generally falling under the umbrella term of *schema matching* or *ontology alignment*. In practice, simple solutions to ontology alignment get us most of the way in real-world domain-specific KGC pipeline, though advanced solutions are mandated if the accuracy requirements are higher than normal or if the domain is particularty difficult. More recently, there has also been some work on schema-free approaches that do not require an alignment between heterogeneous ontologies before executing an ER workflow. The efficacy of these approaches is not fully understood, however, since only a handful of papers have explored its applications.

Additionally, post-processing steps like clustering may also be required *after* the similarity step has finished executing. We mentioned clustering and transitive closure earlier, and these continue to be the most important post-processing steps.

3.6   Related Work: A Brief Review

Entity Resolution has been a research area for almost fifty years, even though the problem has picked up a large amount of steam only in the last couple of decades owing to the growth of the Web. Recall that we had listed four important challenges when we had first described the problem (i.e. automation, heterogeneity, scalability and domain adaptation). Concerning the last challenge (domain adaptation), we note that most ER solutions in research tend to be domain agnostic, although a few are specifically geared for customer (both people and business) names. Many of the best domain-specific ER systems (such as for product names, or for publications) tend to have been developed in industry, and likely required a lot of proprietary training, tuning and model engineering. Scalability efforts have tended to attract the attention of the database community (by way of devising the problem as optimizing 'soft joins') and to a lesser extent, parallel and distributed systems. There is a tradeoff, however, between automation and scalability, as we later discuss.

Our main focus in the discussion herein will be on automated solutions, with some focus on heterogeneity. The reason is that automation continues to be the most important issue, both in the broader AI community, but especially in problems like ER where intensive effort is usually required to achieve good quality.

3.6.1 Automated ER Solutions

Since the early 2000s, machine learning has been actively applied to ER [57]. A machine learning-based ER system could adaptively learn good blocking and similarity functions from both the labeled training data (for supervised approaches), and the unlabeled data (for unsupervised, semi-supervised and clustering-based approaches). On the other hand, systems that use a fixed set of heuristics on all data sources are non-adaptive, and by any pragmatic definition, the issue of automation trivially does not arise.

One of the earliest examples of an adaptive ER system, proposed by Winkler [183], uses a variant of the Expectation Maximization (EM) algorithm [53]. The Fellegi-Sunter model of record linkage is assumed [61]. In this model, candidate entity pairs are partitioned into three classes (matches, non-matches and possible matches) using two decision thresholds. The class of possible matches includes entity pairs that are too ambiguous for the similarity function to resolve into a match or non-match class. Such pairs require clerical review. A Bayesian argument shows that using two decision thresholds is optimal in the sense of minimizing possible matches for preset Type I and II error rates [61].

Unfortunately, Winkler [184] stipulated that the EM algorithm can only be successfully applied to ER if at least five empirical conditions are met. Elmagarmid et al. [57] succinctly list these conditions, some of which are problematic for Linked Data. One such assumption is conditional independence of features. Another is that the match class is well-separated from the non-match class. In systems were EM was considered as a baseline, the empirical performance was found to be less than ideal when some of the stated assumptions are violated [89].

Ravikumar and Cohen [150] use similar, but more robust, ideas by proposing hierarchical graphical models as a way of modeling the similarity of features through latent variables. The system is unsupervised, but assumes structural homogeneity and a serial architecture. A distance function[1] is also assumed to be provided. Empirically, the scope of the work was limited to Relational Database deduplication applications.

On a similar note, Bhattacharya and Getoor [20] use Latent Dirichlet Allocation (LDA) for modeling latent commonalities between entities [29]. The main appli-

---

[1] In the paper, Soft-TF-IDF was proposed as an excellent distance function [150].

cation of their work is in *collective* classification [21]. A classic example arises in the co-authorship domain. Given a set of bibliographic works, two authors (on two independent works) are likely to be the same individual if they have similar co-authors. By modeling such relational information through latent variables, pairs of individuals can be collectively disambiguated. While promising, the work has not been shown to be applicable to domains where relational issues don't arise. Similar to the work by Ravikumar and Cohen [150], structural homogeneity and serial execution were both assumed in the original paper [20].

Christen [39] adopts a different approach. First, a strong weight-based heuristic is used to sample training examples that are almost certainly matches or non-matches. Intuitively, the feature weights in such examples are nearly all 1.0 for matches (or 0.0 for non-matches). The method is predicated on locating such extreme-weighted samples to bootstrap the training process. A classifier (SVM) is trained on the samples and used to label other feature vectors in the candidate set. The method, along with other viable classifiers, a synthetic data generator and a user interface, is available in the FEBRL toolkit [40]. FEBRL was originally designed for biomedical record linkage (a much more constrained form of ER, pertaining primarily to the database community), but can be applied to other domains. Heterogeneity is a major issue, since FEBRL is designed for structurally homogeneous applications. Empirically, only small benchmarks were used for evaluations.

Systems based on active learning have also been proposed, two good examples being RAVEN and COALA [125, 128]. Such systems do not require as many training examples as fully supervised systems such as MARLIN [23], and deliver competitive performance. A major disadvantage is scalability, owing to the method being iterative and requiring continuous user participation. On a positive front, heterogeneity is less of an issue as these systems were designed for Linked Data applications. In particular, RAVEN accommodates structural heterogeneity by modeling type and property alignments as an application of the stable marriage problem [72].

Genetic algorithms have also been extensively explored [126], both in supervised and unsupervised versions. The unsupervised version relies on a measure known as a pseudo F-Measure (PFM). PFMs are heuristics that aim to approximate the actual F-Measure by analyzing the data, and are used as fitness functions in the genetic algorithms. A PFM can also be used to guide the unsupervised learning of a link specification function, as in the deterministic EUCLID algorithm, which uses linear and Boolean classifiers [127]. Although promising, evaluations have shown that the correlation between various proposed PFMs and the actual F-Measure is tenuous [127]. With genetic approaches, the entire dataset has to be scanned over multiple iterations, and results are non-deterministic. In the original papers, EUCLID and the genetic algorithms also did not include solutions for type and property alignments, and were evaluated on small benchmarks [126, 127]. Taken together, these observations indicate that these algorithms may not be suitable for large-scale Linked Data applications.

A promising solution that requires training data, but that can then be applied to other datasets with minimal supervision through transfer learning was proposed by

Rong et al. [159]. This solution is also one of the few to favor both automation and heterogeneity, the latter by virtue of employing schema-free features. An example of a schema-free technique that was earlier introduced in Chap. 2 was Canopies [112]. Such techniques address heterogeneity in a brute-force fashion, by ignoring all structural information. In the case of [159], features are extracted by jointly considering the information set of all properties (of a candidate instance pair). For example, a numeric parser is used to extract numeric information (e.g. dates) present in the properties. A problem with using such features is that noise can be introduced by extracting irrelevant information. Also, Rong et al. [159] do not directly address type heterogeneity. Finally, while transfer learning has some advantages, it also degrades occasional performance. Determining when to use transfer learning is an ongoing area of research [136].

3.6.1.1 The Automation-Scalability Tradeoff

There is a queer tension between automation and scalability. Extremely scalable ER systems, such as Dedoop, require user involvement in terms of specifying the workflow, as described below. Broadly speaking, it has been found that scalable systems tend to make strong assumptions. Locality Sensitive Hashing techniques, for example, assume that appropriate hashing families exist for the distance functions being approximated [52]. In literature covering both ER and ontology or schema matching, the only functions for which LSH has been appropriately utilized are Jaccard and a version of the cosine distance function [56]. An extension to LSH techniques to accommodate the properties of machine learning classifiers is by no means straightforward. Another example of an architecture amenable to parallel and distributed algorithms, Swoosh, also imposes strong assumptions on the similarity function [15].

It is also interesting to note that ER systems implemented in a shared-nothing paradigm, such as MapReduce, tend to leave the burden of specification on the user. We mentioned Dedoop earlier as an example that requires the user to completely specify the workflow [94]. The same is true for LIMES and SILK [124, 172], which are not implemented in MapReduce, but require the user to specify the appropriate functions and parameters. Smart joins and soft joins, which have witnessed much research in the database community, are not adaptive and generalizable in the way proper ER systems are.

A promising approach that is potentially amenable to a fixed number of approximately linear-time MapReduce jobs is the SVM-based proposal by Christen [39]. In its present form, the proposal accommodates neither scalability nor heterogeneity. The latter problem can be dealt with, as described in the following section. It is less obvious how the system can automatically and scalably locate good seed examples to bootstrap the training process. Christen makes the assumption that seeds can be unambiguously located by seeking feature vectors with weights that are nearly all 0 or 1. With noisy data, this is almost never guaranteed. In empirical findings, feature vectors are often found to be sparse, even for duplicates. If a potential method can locate seeds from such data using a fixed number of MapReduce jobs automation and scalability requirements can be reconciled. Once located, seeds can be used, in principle, for training not just a machine learning similarity classifier, but also for learning DNF blocking schemes and determining property alignments.

It is also possible to survey this issue from the opposite end of the spectrum. Automated systems, which mainly tend to be EM-based algorithms that iteratively refine a likelihood function by learning good parameters for latent variables, require multiple scans over the dataset, copious amounts of data sharing and an unspecified number of iterations before convergence [20, 150]. In general, they are non-deterministic and may require multiple re-starts to avoid the pitfalls of local optima. As Winkler (1993) observed [183], various empirical conditions have to hold for such algorithms to be viable. Recent progress on this last issue has been promising, but is, by no means, a settled matter [159].

Generalizing even further, the problem arises because automated systems yield a similarity function that may be 'opaque' in that it is a black-box function that cannot be further analyzed or optimized. In such cases, the only option is to rely on a good blocking function, which may itself require manual intervention. We subsequently describe some efforts in the direction of discovering good blocking functions without manual supervision. There has been some work on this, but by and large, there is no workflow that is both unsupervised and that is ultra-scalable.

3.6.2 Structural Heterogeneity

A traditional assumption in the ER community is that datasets have been homogenized prior to executing an ER workflow [96]. In the tabular community, schema matching is assumed to have been performed a priori [41, 57]. In the Semantic Web community, ontology matching is assumed to have been performed a priori [63].

These assumptions would not be problematic if the schema and ontology matching problems were solved. In fact, research on them has been ongoing for many decades [60, 147]. In some cases, schema matching systems like Dumas assume that ER has been solved a priori[2] [24]. The argument is that, despite the progress in both ER and schema matching, it is misleading to assume that either problem has been solved perfectly.

The question is largely empirical. Is it sufficient to use classic, relatively simple, approaches to address type and property heterogeneity in the broader context of ER? Empirical results have shown that while type heterogeneity is amenable to classic approaches [88], property heterogeneity is not [89]. Insofar as the related work is concerned, only the RAVEN system properly[3] deals with heterogeneity, although empirical evaluations on this issue are limited [125]. Other systems, like the one

---

[2]Dumas relies on duplicates to match columns.

[3]That is, RAVEN addresses heterogeneity through alignments, as opposed to ignoring structure.

by Rong et al. [159], address heterogeneity by using schema-free features that ignore structural properties altogether. There is an empirical argument against such approaches, for well-structured RDF graphs, since one would be losing valuable structural information by adopting a purely schema-free approach in devising features (while simplistic, an analogy would be the process of 'getting rid' of columns in a table by concatenating all columns into one column, which would make the schema matching process trivial, since one would be matching records between tables with only one column each).

Note that the barrier to adopting a heterogeneous solution is conceptually simple to overcome, by pre-pending alignment modules to the basic two-step workflow illustrated earlier. Recent progress on this issue has been promising, especially in the context of blocking [137].

3.6.3   Blocking Without Supervision: Where Do We Stand?

We mentioned earlier that there has traditionally been a strong focus (in the ER research community) on the similarity step. An unfortunate consequence of the complexity of recent ER research is that researchers often ignore other aspects, such as blocking, in their exclusive focus on similarity or scalability. For example, both [150] and [20] use simple ad-hoc blocking keys in their experiments.[4] Scalable systems make more extreme assumptions. For example, Dedoop require both blocking and similarity steps to be precisely specified by a user as part of an ER workflow [94].

For the same reason that ignoring the effects of schema or ontology matching prior to ER is dangerous, the effects of blocking on the overall workflow should not be neglected. Before a contribution in 2013 [87], DNF blocking scheme learners were, at best, semi-supervised [33]. Evaluations conducted in prior work by the author show that clustering techniques such as Canopies do not work well on a variety of datasets [112]. In the Semantic Web, the only unsupervised blocking scheme learner, proposed by Song and Heflin in 2011 [166], was evaluated on small datasets and is not as expressive as general-purpose DNF blocking schemes. Thus, in real-world ER, unsupervised blocking cannot be assumed away, since it is not completely solved yet. Whether the community addresses this issue in sufficient detail, as opposed to a continuing skewed focus on the similarity step, will become clearer in the decade to come.

---

[4]For example, all records sharing a 4-gram character sequence were placed in the same block [150].

3.7 Summary

Entity Resolution is an important second step in a knowledge graph construction workflow following information extraction. The problem continues to be a difficult one, and has taken on renewed importance with the advent of knowledge graph ecosystems. Although the research has witnessed much progress, some issues are still outstanding. Particularly glaring is the lack of an adaptive, unsupervised ER system that learns from prior results, is amenable to domain adaptation and transfer and is scalable enough to deal with Web-scale graphs. Also lacking is a systems-level view of the problem, although in recent years, research has been catching up to industry in building end-to-end ER infrastructures for specific problem domains (such as products and geopolitical events).

Chapter 4
Advanced Topic: Knowledge Graph Completion

4.1 Introduction

Information extraction and entity resolution are clearly both important steps in domain-specific knowledge construction [66, 111]. However, even when done with high accuracy, they are rarely enough. Knowledge graphs constructed over raw data very often have missing and noisy information, including incorrect triples and missing relations. Put simply, such knowledge graphs have to be *refined* or 'completed' before they can be deployed in a good application [139, 145]. An example is illustrated in Fig. 4.1. The knowledge graph fragment in the figure describes political figures and their affiliations, and is possibly extracted from news articles. We assume for the moment that the entities and relations have been correctly extracted and reconciled (i.e. the techniques in Chaps. 2 and 3 achieved excellent performance). Given this KG, if we were to execute a query asking who the first lady was under President Barack Obama's presidency, we would not get any answer. On the other hand, we would get a response from the system if we replaced President Barack Obama in the query with President George Bush. This is because the fact that Laura Bush was the first lady in the Bush administration has been explicitly extracted from a source, perhaps because it was mentioned, while the same is not true for First Lady Michelle Obama. In general, it is not reasonable to assume that every possible fact or inference is ever going to be explicitly extracted from a raw input data source. Sometimes (as in the case above), this is because the 'missing' fact is not mentioned in the source explicitly, but many times, it is also because of noise in the extraction system. Similar reasoning can be applied to the presence of 'wrong' links.

In its broadest form, knowledge graph completion would take a graph with missing and wrong edges and nodes, and attempt to both correct and complete the graph. In other words, knowledge graph correction is included within knowledge graph completion. In the case of Fig. 4.1, a 'good' system would be able to take the graph and infer the fact that Michelle Obama was first lady in the Obama

M. Kejriwal, *Domain-Specific Knowledge Graph Construction*, SpringerBriefs in Computer Science, https://doi.org/10.1007/978-3-030-12375-8_4

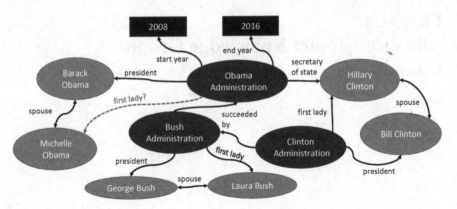

Fig. 4.1 A simplified illustration of the knowledge graph completion problem

administration. A 'bad' system would add noisy links, or remove correct links by incorrectly labeling them as noisy. Usually, the picture is much more nuanced and the evaluation of a knowledge graph completion system involves assessing whether, despite the potential introduction of some noise, the system ended up providing an overall benefit to knowledge graph quality.

Why is there reason to believe that knowledge graph completion works? One can intuitively see that the global graph exhibits some 'semantic regularities' that could be exploited. For example, if we had observed ten presidents in the KG, and found that 90% of their spouses were also explicitly designated first ladies in the KG, it is reasonable to believe that the other 10% are also first ladies, despite no explicit evidence. One can also see why this kind of inference can be a problem. The question of when and where knowledge graph completion is useful, and when it should be avoided has not been fully addressed by the research community. One disadvantage of completion, and of any method that relies solely on inference, is that state-of-the-art neural methods typically no provenance or 'explanation' of why some link was predicted between two nodes, or why some link was declared as noisy.

With these caveats in mind, we argue that knowledge graph completion is still a very useful, and actively researched, area within the broader community of knowledge graph construction. Multiple classes of methods have been proposed for the problem over the years. Before the modern renaissance of deep learning and neural networks, probabilistic graphical models constituted the primary line of attach for such 'relational' problems. Markov logic networks were particularly popular in the mid 2000s, but for various reasons, including scalability, were supplanted by models like probabilistic soft logic that offered a good tradeoff between expressibility, optimization and representation. Probabilistic soft logic and its variants have continued to be popular for various tasks, but the dominant line of research in the knowledge graph completion community (at the time of writing), and the one that we subsequently describe, is *knowledge graph embeddings* (KGEs) [176].

4.2 Knowledge Graph Embeddings

Knowledge Graph Embeddings (KGEs) are an applied offshoot of a broader emergent research area called *representation learning* [14]. In the 2000s, as machine learning systems continued to proliferate, it was realized (somewhat disquietingly) that the performance of a machine learning system was usually dependent very heavily on the *features* engineered over the datasets, as opposed to the virtues of the machine learning algorithms themselves. Feature engineering tended to be manual and ad-hoc, and in the general case, there was no good reason to suppose why one feature would perform better than another. Researchers also realized that the 'goodness' of a feature set also had a lot to do with the dataset itself i.e. it was quite possible for a particular set of features to perform well on one dataset but not another, all else being the same. Clearly, a less manually intensive, data-driven approach to learning good representations for data (whether it was natural language data like words and sentences, or image data) was motivated.

In the natural language community, representation learning algorithms like word2vec have become fairly standard and are widely used across the board for a range of tasks [118], especially information extraction (relation extraction, event extraction and named entity recognition). The key idea is to slide a window over text and to *embed* each word into a dense, real-valued vector space (typically between 50–200 dimensions) that is low-dimensional compared to alternatives like tf-idf, which require dimensionality in the range of 50,000–1 million, depending on the size of the language's vocabulary. The optimization function used for the embedding takes into account the other words in the window, called a *context*. Intuitively, words that have similar context would be embedded close together in the vector space. In natural language, this generally leads to common-concept instances (such as cat and dog) being embedded close together due to their similar context. This kind of embedding is reminiscent of topic models like Latent Dirichlet Allocation (LDA) [29], but LDA was a graphical (not neural) model designed to embed *documents*. In contrast, algorithms like word2vec are designed to embed words, based on context, rather than coarser-grained documents, although variants of word2vec can also be used to embed sentences, paragraphs and documents [51].

Because of their semantic dependence on context, rather than ontology, embeddings based on statistical models have been found to capture some remarkable analogical patterns in a completely unsupervised fashion (Fig. 4.2). Despite not being given ontological information, the embedding is able to deduce that words like 'generator' and 'battery' should be clustered closer together in a semantic space, rather than (say) 'teaching' and 'generator'. In the embedding space, one can also carry out vector operations like **King − Man + Woman** with the resulting vector being very close to **Queen** in the semantic space.

Primarily because of these semantic properties, and also (on a related note) because of superior performance on downstream natural language processing tasks like information extraction, the word2vec algorithm became so popular that numerous variants have emerged, and the algorithm has even been adopted to embed

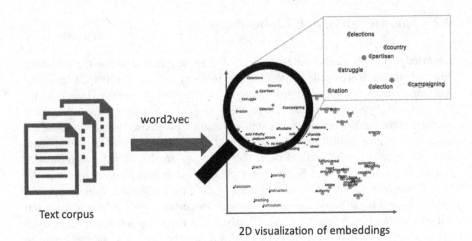

Fig. 4.2 An illustration of word embeddings using algorithms like word2vec. Words that occur in similar contexts (e.g., elections, campaigning) are clustered closer together in the vector space. 2D visualizations like these (from higher-dimensional vectors) can be rendered using neural dimensionality reduction algorithms like t-SNE [108]. Note that dimensions do not have any intrinsic meaning

nodes in networks and graphs (see e.g., the DeepWalk algorithm [141]). However, its application or adoption to knowledge graphs is not clear, and has not been usefully demonstrated. In part, the reason is that knowledge graphs are a richer, more structured data set, since even the simplest definition of a KG assumes that is a multi-relational, directed, labeled graph where entities are nodes and relations are different types of edges. Motived by this additional structure, novel approaches were proposed in the early 2010s to embed nodes (and often, but not always, relations) in the KG into a continuous vector space while preserving certain key structural properties.

In the rest of this section, it will be useful to think of a KG as a set of triples, where each triple is of the form (h, r, t), where h and t are referred to as head and tail entities respectively, and r is the relation. We do not assume constraints, although models like RDF impose many requirements on how relations and head/tail entities are actually represented. For example, the RDF model does not allow head entities to be modeled as 'literals' like strings or numbers. One reason for ignoring such constraints in the present discussion is that typical (and early) papers on knowledge graph embeddings have mostly arisen in the NLP and general AI communities, rather than the Semantic Web community, which is the major adopter of RDF. Although some recent work has attempted to embed knowledge graphs modeled specifically as RDF, even these models tend to be heavily inspired by the earlier models that were proposed in 'ontologically light' communities.

Furthermore, although many embedding models exist at the time of writing, almost all models represent h and t as points in the vector space, while relation r usually has a more flexible representation, since it is modeled as an *operation*. Thus, it could be a vector, representing operations such as translation or projection, but in some cases, it can also be a matrix. In contrast, representing an entity as a matrix is far less common. The representations themselves are learned by minimizing a global loss function involving all entities and relations. As a result, even the embedding representation of a *single* entity or relation encodes global information from the *entire* knowledge graph. Subsequently, we describe some of the more established methods for achieving this kind of encoding.

4.2.1 TransE

TransE was one of the first KGE techniques proposed (shortly after the *Structured Embedding* method [31]) [30], and has largely survived the test of time. It continues to be widely used, and delivers extremely competitive performance. A range of alternatives (commonly referred to as Trans*) have been built using the same fundamental principles as TransE, but with richer optimizations and information sets. We present the TransE system in detail, and then briefly cover some alternatives.

First, TransE is an *energy-based* model for learning low-dimensional embeddings of entities; specifically, relationships are represented as translations in the embedding space: if (h, r, t) holds, then the embedding of the tail entity t should be close to the embedding of the head entity h plus some vector that depends on the relationship r. Put more mathematically, the algorithm attempts to generate an embedding for each h, r and t such that for triples observed in the knowledge graph, the *translation* relationship $\mathbf{h} + \mathbf{r} \approx \mathbf{t}$ should hold. Given enough triples, the hope is that the embedding is general enough to yield new information i.e. in the test phase, if we observe a relationship $\mathbf{h}' + \mathbf{r}' \approx \mathbf{t}'$ that was not observed during training, there is a non-trivial probability that (h', r', t') is a *true* triple (constituting missing information in the original KG) and can be added to the KG to complete it. A key strength of TransE is its reliance on a reduced set of parameters since it learns only *one* low-dimensional vector for *each* entity and *each* relationship. The energy-based optimization function (based on translation) is also simple and intuitive to understand.

Why would translation be expected to be so successful? One motivation is that hierarchical relationships are extremely common in KBs and translations are the natural transformations for representing them. For example, a natural representation of trees is to have the siblings be close to each other; in other words, with nodes at a given height organized on the x-axis, the parent-child relationship corresponds to a translation on the y-axis. Since a null translation vector corresponds to an equivalence relationship between entities, the model can then represent the sibling relationship as well. A secondary motivation arose from coincidental findings from the word embedding literature, where some authors found that many relations (e.g., capital-of, has-father) are represented by the model as translations in the embedding space. This suggested the existence of embedding spaces in which 1-to-1 relationships between entities of different types may potentially be represented by translations. The intent of TransE was to enforce such a structure of the embedding space.

4.2.2 TransE Extensions and Alternatives

The basic TransE model has been extended in numerous ways into a family
of Trans* algorithms such as TransH, TransR, and TransD, to name a few
[106, 176, 179]. The primary difference lies in the underlying assumptions about
the embedding space,, which impacts the optimization function used during both
training and testing. For example, to overcome limitations of TransE in dealing with
1-to-N, N-to-1, and N-to-N relations, an effective strategy is to allow an entity to
have *distinct* representations when involved in *distinct* relations. To take an example,
one could imagine learning a different embedding for the city 'Tokyo' in the context
of a relation such as 'capital of' than in the context of the relation 'has population'.
Intuitively, the first embedding would put (the entity vector representation of) Tokyo
closer to other capital cities, while the second embedding may place it closer to
cities with similar populations. in theory, this kind of embedding permits richer
information sets to be captured, but at the cost of using more training data and
observing more triples.

TransH follows this general idea [179], by introducing *relation-specific hyperplanes*. Similar to TransE, TransH models entities and relations as vectors, but
the relation vector r is considered to be specific to a hyperplane (defined by its
normal vector w_r). In other words, a relation is actually captured by two pieces
of information (a vector, and a hyperplane normal). Given a triple (h, r, t), the
entity representations of h and t are first *projected* onto the hyperplane, followed
by the translation operation. The projections are assumed to be connected by r on
the hyperplane with low error if (h, r, t) holds, with the scoring function being
similar to that used by TransE. It is both possible and expected that, for some
hyperplanes, the triple will have low error, while on other hyperplanes it won't,
since h and t will not be connected through the relation underlying that hyperplane.
Mathematically, the optimization is richer since it has to perform the dual task of
hyperplane-specific translation (low error for true triples and high error for false
triples), and discovering hyperplanes that are expressive and separable enough to
accomplish such discrimination. As a consequence, TransH is slower than TransE,
and (all else being equal) does not generalize as well to smaller or sparser graphs
than TransE.

TransR is a similar variant [106], the difference being that it introduces relation-
specific *spaces*, rather than (the more constrained) *hyperplanes*. In TransR, entities
are represented as vectors in an entity space $\mathscr{R}^d$, and each relation is modeled as a
translation vector in k-dimensional space $\mathscr{R}^k$ that doesn't necessarily have to be a
hyperplane. More details on these operations are provided in the original paper.
Herein, we note that, although powerful in modeling complex relations, TransR
introduces a projection matrix for *each* relation, hence requiring $O(dk)$ parameters
per relation. It ends up losing both the simplicity and efficiency of TransE and
TransH, both of which require only $O(d)$ parameters per relation, d being the
embedding dimensionality. More complicated versions of the same approach were
also later proposed. We do not cover these here, but provide a list of some models
(and their embedding spaces) published at the time of writing, in Fig. 4.3.

Method	Ent. embedding	Rel. embedding
TransE	$\mathbf{h}, \mathbf{t} \in \mathbb{R}^d$	$\mathbf{r} \in \mathbb{R}^d$
TransH	$\mathbf{h}, \mathbf{t} \in \mathbb{R}^d$	$\mathbf{r}, \mathbf{w}_r \in \mathbb{R}^d$
TransR	$\mathbf{h}, \mathbf{t} \in \mathbb{R}^d$	$\mathbf{r} \in \mathbb{R}^k, \mathbf{M}_r \in \mathbb{R}^{k \times d}$
TransD	$\mathbf{h}, \mathbf{w}_h \in \mathbb{R}^d$ $\mathbf{t}, \mathbf{w}_t \in \mathbb{R}^d$	$\mathbf{r}, \mathbf{w}_r \in \mathbb{R}^k$
TranSparse	$\mathbf{h}, \mathbf{t} \in \mathbb{R}^d$	$\mathbf{r} \in \mathbb{R}^k, \mathbf{M}_r(\theta_r) \in \mathbb{R}^{k \times d}$ $\mathbf{M}_r^1(\theta_r^1), \mathbf{M}_r^2(\theta_r^2) \in \mathbb{R}^{k \times d}$
TransM	$\mathbf{h}, \mathbf{t} \in \mathbb{R}^d$	$\mathbf{r} \in \mathbb{R}^d$
ManifoldE	$\mathbf{h}, \mathbf{t} \in \mathbb{R}^d$	$\mathbf{r} \in \mathbb{R}^d$
TransF	$\mathbf{h}, \mathbf{t} \in \mathbb{R}^d$	$\mathbf{r} \in \mathbb{R}^d$
TransA	$\mathbf{h}, \mathbf{t} \in \mathbb{R}^d$	$\mathbf{r} \in \mathbb{R}^d, \mathbf{M}_r \in \mathbb{R}^{d \times d}$
KG2E	$\mathbf{h} \sim \mathcal{N}(\mu_h, \Sigma_h)$ $\mathbf{t} \sim \mathcal{N}(\mu_t, \Sigma_t)$ $\mu_h, \mu_t \in \mathbb{R}^d$ $\Sigma_h, \Sigma_t \in \mathbb{R}^{d \times d}$	$\mathbf{r} \sim \mathcal{N}(\mu_r, \Sigma_r)$ $\mu_r \in \mathbb{R}^d, \Sigma_r \in \mathbb{R}^{d \times d}$
TransG	$\mathbf{h} \sim \mathcal{N}(\mu_h, \sigma_h^2 \mathbf{I})$ $\mathbf{t} \sim \mathcal{N}(\mu_t, \sigma_t^2 \mathbf{I})$ $\mu_h, \mu_t \in \mathbb{R}^d$	$\mu_r^i \sim \mathcal{N}(\mu_t - \mu_h, (\sigma_h^2 + \sigma_t^2)\mathbf{I})$ $\mathbf{r} = \sum_i \pi_r^i \mu_r^i \in \mathbb{R}^d$
UM	$\mathbf{h}, \mathbf{t} \in \mathbb{R}^d$	—
SE	$\mathbf{h}, \mathbf{t} \in \mathbb{R}^d$	$\mathbf{M}_r^1, \mathbf{M}_r^2 \in \mathbb{R}^{d \times d}$

Fig. 4.3 Embedding spaces of TransE and its alternatives. $\mathcal{N}$ is the normal distribution with the usual mean μ and standard deviation σ parameters. d is the embedding dimensionality (set by the user, and generally in the range of tens to hundreds), n and m are the numbers of entities and relations respectively in the KG to be embedded. Other symbols are algorithm- or system-specific, although some (such as k) can be specified by the user. For example, in TranSparse θ is the average sparseness degree of projection matrices. For more formal definitions of parameters, we refer the reader to the individual papers or to a recent condensed survey

Method	Space complexity	Time complexity
TransE	$\mathcal{O}(nd + md)$	$\mathcal{O}(d)$
TransH	$\mathcal{O}(nd + md)$	$\mathcal{O}(d)$
TransR	$\mathcal{O}(nd + mdk)$	$\mathcal{O}(dk)$
TransD	$\mathcal{O}(nd + mk)$	$\mathcal{O}(\max(d, k))$
TranSparse	$\mathcal{O}(nd + (1 - \theta)mdk)$	$\mathcal{O}(dk)$
TransM	$\mathcal{O}(nd + md)$	$\mathcal{O}(d)$
ManifoldE	$\mathcal{O}(nd + md)$	$\mathcal{O}(d)$
TransF	$\mathcal{O}(nd + md)$	$\mathcal{O}(d)$
TransA	$\mathcal{O}(nd + md^2)$	$\mathcal{O}(d^2)$
KG2E	$\mathcal{O}(nd + md)$	$\mathcal{O}(d)$
TransG	$\mathcal{O}(nd + mdc)$	$\mathcal{O}(dc)$
UM	$\mathcal{O}(nd)$	$\mathcal{O}(d)$
SE	$\mathcal{O}(nd + md^2)$	$\mathcal{O}(d^2)$

Fig. 4.4 Time and space complexity of selected KGE models. For details on notation, see caption of Fig. 4.3

4.2.3 Limitations and Alternatives

Recently, it was shown that KGEs can suffer from problems of generalization, reflected in poor performance, when the KG is either too sparse or noisy (or both) [144]. In such situations, alternate, more established methods such as Probabilistic Soft Logic (PSL) were found to work better [92]. Another issue is the time taken to train an embedding, and the tuning of hyperparameters. While an efficient implementation of TransE (and some of its extensions) exists at the time of writing, the original implementation was time consuming requiring on the order of hours to train medium-sized knowledge graphs. This makes trial-and-error-style training and tuning, problematic. Over time, the models have become steadily more complicated, in fact (Fig. 4.4).

4.2.4 Research Frontiers and Recent Work

Many of the models that have been proposed for knowledge graph embeddings are based on using assertions in a given KG as observations in the training data. However, this imposes a degree of locality on the embedding model, since there are other potential information sets that can be considered in the embedding optimization. Some possibilities (as alternate information sets) that have been recently proposed in addition to, or even instead of, assertions in the KG are covered below. These information sets can be used to augment and improve KGE training [176].

4.2.4.1 Ontological Information

Knowledge graphs do not just contain entities, since (as we described in detail in both Chaps. 1 and 2) many of the nodes are *typed* according to some ontology, whether implicit, explicit or shallow. Some ontologies, such as YAGO [167], can be extremely detailed containing full hierarchies of classes and sub-classes. For example, Sharon Stone is a person, but also an actor. Thus, all else the same, we would give higher weight to a triple that declares Sharon Stone to have starred in a movie, than if we had not known that Sharon Stone is an actress. Many similar examples can be formulated along these lines.

Modeling this intuition in a computational way is less straightforward. A possible avenue is to *not* treat the 'is-a' relation as special (recall that is-a was one of the few, and often the only, relations that serves as a 'glue' between the KG and the ontology) but to declare all is-a triples, and possibly other ontological triples, in the same vein as other KG triples e.g., $< SharonStone, is-a, Actor >$, $< SharonStone, is-a, Person >$. In this formulation, we are essentially *augmenting* the original training set (the assertions in the KG) with additional triples (the ontology, and the glue between KG and ontology). The effectiveness of this strategy is not completely clear at the time of writing, especially with respect to the relative sizes of ontologies and KGs. An advantage of the method is that it is simple. A disadvantage is that it may be simplistic e.g., is Sharon Stone is an actress, an actress is an entertainer, and an entertainer is a person, the embedding is not really capturing the fact that Sharon Stone is a person. Intuitively, the *special semantics* of is-a (and other) relations is not being taken into account by the embedding.

This has motivated more complex approaches that take into account the special nature of is-a triples. For example, as proposed in [70] using a method called *semantically smooth embedding* (SSE), one could *explicitly* design the optimization to require entities of the *same type* to stay close to each other in the embedding space e.g., Sharon Stone would be closer to Sylvester Stallone than to Roger Penrose, since Stallone is also an actor, while Penrose is a scientist. Technically, SSE employs two manifold learning algorithms, i.e., Laplacian eigenmaps and locally linear embedding to model such a 'smoothness assumption'. More specific details can be found in the original paper [70].

A second approach, proposed in [186], is *type-embodied knowledge representation learning* (TKRL), which can handle hierarchical entity categories and multiple category labels. TKRL is a translational distance model with type-specific entity *projections*. Given an assertion (h, r, t), TKRL first projects h and t with type-specific projection matrices, and then models r as a translation between the two projected entities. Because of the matrices, TKRL can have a high space complexity, and would likely not generalize well unless it had access to enough data. When it does, however, it has been found to have better performance in tasks and applications such as link prediction and triples classification. Whether this tradeoff makes sense for an application designer depends on the application and the size (and trustworthiness) of the KG to be embedded.

Currently, there continues to be ongoing interest in utilizing ontological information sets in knowledge graph embeddings, whether directly or indirectly. Most
likely, there is still much work to be done in this area. What is clear, however, is
that this information should not be ignored by the embedding model and can serve
a useful purpose, whether in terms of boosting performance or (more abstractly)
modeling human intuition more closely.

4.2.4.2   Text

Researchers have also explored incorporating textual descriptions of entities into
the KGE model. This is motivated by the observation that, in many KGs, concise
descriptions for entities are available, containing important semantic information
about the entities. Even when this is not the case, one can always find and crawl
sources such as news releases and Wikipedia articles to enrich entities with context.

Embedding KGs with text information dates back to the NTN model [165], which
was proposed fairly early on. In NTN, text information is used in a fairly naive way
since it is simply used to *initialize* entity embeddings. Namely, NTN first learns
word vectors from a news corpus, and then initializes the representation of each
entity by *averaging* the vectors of words contained in the entity's label. By way of
example, the representation for 'Sharon Stone' would be initialized by averaging
the word vectors for 'Sharon' and 'Stone'. This example also shows why the utility
of text information is naive in this model, since Sharon and Stone individually show
up in other contexts as well. This is also true for locations ('New York' vs. 'New
Orleans', or 'Los Angeles' vs. 'Los Alomos'), and for other entity types as well.
Later, another similar method was proposed, in which entities were represented
as average word vectors of their descriptions rather than just their names. More
generally, these kinds of methods are problematic because they do not take into
account *joint contexts* of assertions and text but instead model textual information
distinctly from assertions, and in the process, fail to leverage the potentially rich
interactions between such information sets.

To the best of our knowledge, the first such joint model was proposed in [178].
The main idea was to *align* the KG with the text corpus, and then train both KG
embedding and word embedding jointly, with the hope that both embeddings will
be informed and improved by each other since the embeddings for entities, relations
and words are all represented in the *same* vector space. Operations like inner
product and similarity between these different elements become meaningful and
insightful. The joint model has three 'sub-models': knowledge, text, and alignment.
The knowledge sub-model simply embeds entities and relations in the KG and is
actually a variant of TransE, with a special loss function for measuring fitness of the
embeddings to KG facts. The text sub-model embeds words, and is a variant of the
famous skip-gram word embedding model. Similar to the knowledge sub-model, it
comes with a loss function that measures the fitness of the sub-model embedding
to co-occurring word pairs. Finally, the alignment sub-model is designed to ensure
that the embeddings of the two other sub-models lie in the *same space*. Different

such alignment mechanisms are introduced in their work and others that followed it, including by entity names, Wikipedia anchors, entity descriptions etc. with more such mechanisms continuing to be proposed in current research. Similar to the other sub-models, the alignment sub-model's loss function is defined to measure the quality ('fitness') of alignment. We highlight that one of the major contributions in using such joint models is the prediction of *out-of-KG* entities, i.e., phrases appearing in web text but not included in the KG yet.

Yet another approach along these lines is the *description-embodied knowledge representation learning* (DKRL) [185], which seeks to extend TransE to incorporate entity descriptions. DKRL associates each entity with two vector representations, one of which is structure-based and the other of which is description-based. The former captures structural information (just like the original TransE), while the latter captures textual information in the entity description. The description-based representation is constructed using word embeddings. Entity, relation, and word embeddings can all be learned simultaneously by minimizing a ranking loss function when training. Experimental results demonstrated the superiority of DKRL over TransE, particularly in the zero-shot scenario with out-of-KG entities.

Generally, incorporating text into the optimization tends to lead to empirical improvements. However, we are not aware of a full-scale empirical study that attempts to measure the extent of these improvements, and to assess the sensitivity of such improvements with respect to important parameters such as the size and quality of a KG, the relevance of the text corpus, and the noise in the text corpus. Beyond normal performance benefits, a primary benefit of the joint model, as we highlighted earlier, was its ability to gracefully handle entities that may not have been observed in the actual KG.

4.2.4.3 Other Extrinsic Information Sets

Incorporating text and ontological information into KGEs continue to be important directions of research, especially for improving KGEs using more context. However, these information sets are by no means the only ones. Below, we briefly cover some others.

Rules Ontologies are not just sets of inter-related concepts and properties. They can also contain constraints and rules to further express the domain. Can rules, as understood in this sense, be used to further influence KGEs in a positive direction? Wang et al. [177] proposed an approach utilizing rules to refine embedding models during KG completion. In their work, KG completion is formulated as an ILP (integer linear programming) problem. Specifically, the objective function is generated from embedding models, and the ILP constraints from pre-specified rules. Assertions inferred in this way will be the most preferred by the embedding models but would also comply with all the rules. A similar work that combines rules and embedding models via graphical models such as Markov logic networks was later introduced in [181]. However, in both the papers above, rules are modeled separately from embedding models, serving as *post-processing* modules. They do not directly influence embeddings, and hence cannot be used to obtain 'better' embeddings.

A later approach that tried to influence embeddings in a joint model that leveraged rules directly in the embedding optimization was KALE [71], wherein a model was proposed that simultaneously embeds assertions and rules. In this framework, an assertion was modeled as a ground atom, with a well-defined truth value. Also, logical rules are first instantiated into ground rules, with ground rules then interpreted as complex formulae constructed by combining ground atoms with logical connectives (e.g., $\vee$), and modeled by t-norm fuzzy logics [93]. The truth value of a ground rule is a composition of the truth values of the constituent ground atoms, via specific t-norm based logical connectives.

The values of these connectives lie within the range of [0,1], indicating to what degree the ground rule is satisfied. In this way, KALE represents facts and rules in a unified framework, as atomic and complex formulae respectively. After unifying assertions and rules in this way, KALE minimizes a global loss involving both to learn both entity and relation embeddings. These learned embeddings are compatible not only with assertions in the training corpus but also with rules, which is hoped to lead to better performance of embeddings in downstream applications.

KALE has inspired other variants. For example, in [157], the overall approach is similar to KALE, but vector embeddings are introduced for entity pairs rather than individual entities, making it particularly useful for *relation extraction*. This is an example of an embedding which is (a priori) optimized keeping a target application in mind. However, since entities do not have their own embeddings, relations between unpaired entities cannot be effectively discovered.

Both KALE and the variant described above have the limitation that they have to instantiate universally quantified rules into ground rules before learning the embedding models. This grounding procedure is known to be time- and space-inefficient, especially when there are many entities in the KG or the rules are too complex. Some recent work has recognized this drawback, and proposed solutions to address it.

Generally speaking, the ongoing research shows that rules will continue to find more applications and uses in KGEs. The good performance of rule-supplemented KGEs, and the researchers investing in this approach, both show that there is an interesting synergy to be had between methods that were traditionally seen as disparate (statistical and logical), though by no means incompatible. Future research will show till what extent this synergy can be exploited, both in KGEs and other similar areas.

Temporal information In [82], the critical observation was made that KG assertions may often be time-sensitive, e.g., (Sharon Stone, ReceivedAward, Golden Globe) happened in 1995. Based on this observation, a *time-aware embedding* model was proposed, the idea being to impose temporal order constraints on time-sensitive relation pairs, e.g., StarredIn and ReceivedAward. Given such a pair (r_i, r_j), the prior relation is supposed to lie close to the subsequent relation after a temporal transition, i.e., $Mr_i \approx r_j$ where M is a transition matrix capturing the temporal order information between relations. After imposing such constraints, the authors in [82] are able to learn temporally consistent relation embeddings. In other work, [58] tried to model the temporal *evolution* of KGs. That is, changes in a KG always arrive as events, represented by labeled quadruples such as $(h, r, t, s; True)$ or $(h, r, t, s; False)$, indicating that the assertion (h, r, t) appears or vanishes at time s, respectively. Each quadruple is then modeled as a four-way interaction among h, r, t, and s, where s is the vector representation of the time stamp. This model was shown to perform well in dynamic domains such as sensors or medicine. Overall, research has continued to intensify in this domain, and the link prediction problem that we study later has been extended to temporal link prediction i.e. the problem of predicting not just a link, but when it becomes stale (or active).

Paths and structures Relation paths may be understood as *multi-hop relationships* between entities. A relation path is typically defined as a sequence of relations $r_1 \rightarrow \dots \rightarrow r_l$ through which two entities can be connected on the graph. For example, StarredIn $\rightarrow$ ShotIn is a path linking Sharon Stone to Nevada, via an intermediate movie node such as Casino. Relation paths contain semantic cues not otherwise found in the node itself and can useful for KG completion.

More generally, it is possible to learn such 'graph-aware embedding models' by leveraging different types of graph structures. In [62], such a model was proposed, leveraging three types of graph structures: neighbor context (equivalent to triples observed in a KG), path context (similar to relation paths just described) and edge context (defined as the relations linking to and from that entity). The last is primarily ontological e.g., the edge context of Sharon Stone might include relations such as StarredIn, LivesIn, ReceivedAward etc. Intuitively, all of these relations indicate collectively that Sharon Stone is a person, and more specifically, an entertainer. Experimental results have demonstrated the effectiveness of incorporating these graph structures in an embedding model. In other work, [83] suggested that the plausibility of a triple $tr = (h, r, t)$ could be estimated from its immediate context, defined as the set of triples sharing the same head as tr, the set of triples sharing the same tail, the set of triples with h as tail, the set of triples with t as head, and triples with arbitrary relations but where the two entities are h and t. By using such contexts, a system was found to perform better at the link prediction task (described subsequently) on multi-relational data, such as KGs.

Other Entity Attributes When introducing KGs in Chap. 1, we argued that relations in KGs can indicate both relationships *between* entities (e.g., StarredIn indicates a relationship between Movie and Actor entities) or be used to define entity attributes (e.g., the gender or birthdate of a person). Unfortunately, most KG embedding techniques such as TransE do not explicitly distinguish between these semantics. In [131], this distinction was made. Namely, entity-entity relations were encoded in a tensor, while entity-attribute relations in a separate entity-attribute matrix. The matrix and tensor are jointly factorized to learn representations simultaneously for entities, and both types of relations. Similar ideas have since been studied by other authors [105].

4.2.5   Applications of KGEs

Since a knowledge graph embedding is essentially just a mapping from nodes and edges to real-valued vectors, how can we tell when one embedding is better than another? An uncontroversial approach is an *ablation-style evaluation* [182] where, in the context of a given application, we evaluate an embedding against another embedding keeping all else constant (including datasets and metrics). Although such an evaluation is not without flaws, the biases (particularly, dataset bias [170]) are reduced if the benchmarks are large and real-world, and the applications have relevance. Below, we describe several viable applications. Note that one such application, Entity Resolution (ER) [66], has already been covered in detail in the previous chapter. There are two contexts in which we can use KGEs for ER. First, recall that there was a feature generation step whereby we attempted to extract a feature vector for each pair of entities that were consumed in the similarity step of ER. Feature engineering is a labor intensive process and there is always the possibility of missing something. By concatenating the embedding vectors of the two nodes, we can get an alternate feature representation that could potentially be used for better ER performance. Early results have been promising, though the hypothetical utility of embeddings over engineered features is still in the preliminary stages of ER research. A second possibility for utilizing embeddings for ER is to frame ER as a special, supervised case of a relation or link prediction problem (described below).

ER is a good example of an *in-KG* application, which is conducted *within the scope* of the KG where entity and relation embeddings are learned. Three other examples of in-KG applications are link prediction, triple classification and entity classification [176], all of which have been well-studied in the literature. It is not atypical to assume that all of these applications can be cast as special cases of knowledge graph *refinement*, with different definitions of refinement.

Link Prediction   *Link prediction* is the problem of predicting whether a given entity has a specific relation with another 'hypothetical' entity, i.e., predicting h given (r, t) or conversely, t given (h, r), with the former task denoted as head entity prediction $(?, r, t)$, and the latter as tail entity prediction $(h, r, ?)$. Link prediction is a general problem that can be 'fed into' multiple out-of-KG applications e.g., question answering or even conversational AI [65, 168]. For example, $(?, StarringIn, Terminator)$ is to predict the stars of the film Terminator, while $(Sharon Stone, StarringIn, ?)$ amounts to predicting films that Sharon Stone has starred in. This example also shows that prediction can be a many-many problem i.e. there are multiple correct predictions for both cases. Link prediction is a quintessential KG completion task, i.e., adding missing knowledge to the graph, and has been tested extensively in previous literature. An alternate name for the problem (among others) is entity ranking. A similar idea can also be used to predict relations between two given entities, i.e., $(h, ?, t)$, which is usually referred to as relation prediction. In the social network community, link prediction has a much more specific meaning than in the KGE community; it is usually the problem of predicting future links (e.g., friendship) that might be formed between actors in the social network [110].

With entity and relation representations learned during training, link prediction can be done using ranking, similar to procedures developed over decades in the Information Retrieval community. Take the prediction task $(?, r, t)$ as an example, a ranking system can 'predict' the head entity by taking every entity h' in the KG as a *candidate* answer and calculating a score for each (h', r, t), using a scoring function. In descending order of scores, this yields a ranked list of candidate answers. If the embedding is 'good', the hope is that correct predictions will be ranked nearer to the top of the list than incorrect predictions.

Similar to IR, the common way to evaluate such rankings is to use metrics such as *mean rank* (the average of predicted ranks), *mean reciprocal rank* (the average of reciprocal ranks), *Hits@n* (the proportion of ranks no larger than n), and *AUC-PR* (the area under the precision-recall curve). Different metrics have different tradeoffs. AUC-PR takes a balanced view of precision and recall, for example, while Hitsn is oriented more towards recall than precision. For example, considering the Hits10 metric, and assuming there is only one correct prediction, a ranking where the correct entity is at rank 1 will have the same Hits10 ($=1.0$) as one where the correct entity is at rank 10. Similarly, if the correct entity is not in the top 10, but is at rank 11 vs. rank 100, both would receive a Hits10 of 0.0. Note that, for individual 'queries', Hits10 can only be 0 or 1, but when averaged over many such queries, ranges from 0 to 1 and can be used to assess the performance of an embedding on a test set, on average.

Entity Classification *Entity classification* is the problem of classing entities under different *semantic* categories, e.g., Sharon Stone is an Actor, Terminator is a Movie and so on [129]. Generally, the relation that is considered for classification purposes is the *is-a* relation. If the is-a relation has already been included in the embedding process (so an embedding for the is-a relation exists after training), entity classification can simply be treated as a special case of link prediction, and the same evaluation procedures can be applied for it. This similarity between link prediction, and both entity classification and entity resolution, highlights what was noted earlier, namely, that they can all be thought of as very specialized cases of the broader knowledge graph refinement (or completion) problem.

Triple Classification *Triple classification* can be thought of as a binary classification problem [129]: given an arbitrary triple (h, r, t), is the triple *true*? A trivial case is when the triple belongs in the training set, in which case it is clearly true. If this is not the case, then it is not necessary that the triple is untrue, since the training data was incomplete to begin with. One non-trivial issue with framing triple classification as binary classification is the consistent combination of the individual head, tail and relation embeddings in a way that can be used to predict the probability of truth. In systems like the Trans* KGEs (but also others), a density function is used to make such a prediction. The correct metric to use is accuracy, if the test data is balanced. In most benchmarks that have been released so far for triple classification, this has been the case. If the evaluation data is skewed, computing precision, recall or ROC curves may be more appropriate.

We conclude by noting that there are also *out-of-KG applications*, which are generally less controlled since they are designed to scale to broader domains. Such applications include relation extraction, question answering and recommender systems. We do not explore out-of-KG applications in this book, but focus on in-KG applications. More details on out-of-KG applications can be found in an overview of KGEs in [176].

4.3 Summary

Knowledge Graph Embeddings (KGEs) are a powerful set of techniques for representing entities, relations and even descriptions in a KG in a continuous real-valued vector space. Although some of the reasoning capabilities permitted by symbolic representations are lost in the process, the real-valued representations are much less brittle than discrete symbols, and hence, more robust to noise and missing information. Furthermore, recent efforts in the field have tried, with varying success, to reconcile the benefits of continuous and discrete KG representations. Research in this area is still ongoing, but it has become clear that KGEs are vital for the broader problem of knowledge graph completion (or identification). Other applications of KGEs include link prediction and entity classification.

Chapter 5
Ecosystems

5.1 Introduction

Much of this book has described knowledge graphs and their construction at a fairly technical level. In the introduction, we argued that domain-specific knowledge graphs have started to come into their own, using examples such as publications and academia, products and e-commerce, and social causes such as disaster relief. In this chapter, we take a much broader view of knowledge graphs and their impact. Specifically, we attempt an answer to questions such as, *how high has adoption of knowledge graphs been, and in what contexts? What bodes for the future of knowledge graphs?* Although there is a lot more still to come in knowledge graph research, some crystallization has occurred over the last few years (in some cases, decades), which will be the focus of this chapter.

5.2 Web of Linked Data

The Web has provided inarguable benefits but until recently, the same principles that enabled the Web of *documents* to emerge and succeed have not been applied to a hypothetical Web of *data*. Traditionally, data published on the Web was made available as raw CSV or XML dumps, marked up as HTML tables, thereby sacrificing much of its structure and semantics, or in other structured or semi-structured formats that were not intuitive for humans to read or understand (in their raw form). On the conventional Web, driven by hypertext, the *nature* of the relationship between two linked documents is implicit, as the data format, i.e. HTML, is not sufficiently expressive to enable richer semantics and modalities e.g., determining that individual entities described in a particular document are connected by typed links to related entities in the same (or other) documents. In the real world, in contrast, such relationships (between entities) form the basis for knowledge and interaction. A guiding question has been, *can knowledge graphs provide the backbone for enabling such rich, real-world like semantics on the Web?*

There are multiple strands of evidence to indicate that the answer may be yes, although the power of knowledge graphs in enabling semantics on the Web is not limitless [18, 163]. Nevertheless, knowledge graphs published in recent years have contributed greatly to an evolution of the Web from a global information space of linked documents to one where both documents and data are interlinked. Underpinning this evolution is a set of *best practices*, called *Linked Data* [25, 26, 75], for publishing and connecting structured data on the Web. The adoption of the Linked Data best practices has lead to the extension of the Web with a global data space connecting data from diverse domains such as people, books, scientific publications, music, proteins, drugs, statistical and scientific data, and reviews to only name a few. Such a Web enables new application types. There are generic Linked Data browsers which allow users to start browsing in *one* data source and then navigate along links into *related* data sources, analogous to how one could start on an HTML webpage on the conventional Web and then use it to browse to completely different webpages, hosted on servers across the world. There are Linked Data search engines that crawl the Web of Data by following links between data sources and provide expressive query capabilities over aggregated data, similar to queries posed over databases. The Web of Data also opens up new possibilities for *domain-specific* applications. Unlike Web 2.0 mashups, which work against a fixed set of data sources, Linked Data applications operate on top of an unbound, global data space. This enables them to adapt and deliver more complete answers as new data sources appear on the Web.

In its simplest form, Linked Data is about using the Web to create typed links between multi-source data elements such as concepts, entities and properties. These multiple sources may be as diverse as databases maintained by two organizations in different geographical locations, or simply heterogeneous systems within a single umbrella organization that have not traditionally been interoperable at the data level because of problems such as varying schemas, data types etc. Technically, Linked Data refers to data published on the Web in such a way that it is not only machine-readable, but its meaning is explicitly defined ('semantics') [163], it is linked to other external data sets, and can be linked *to* from external data sets.

While the primary units of the hypertext Web are HTML (HyperText Markup Language) documents connected by *untyped* hyperlinks, Linked Data relies on RDF (Resource Description Framework) documents [135]. However, rather than simply connecting these documents, Linked Data uses RDF to make typed statements that link arbitrary things in the world. The result, referred to as the Web of Data throughout this chapter, may more accurately be cast as a Web of *things* in the world, *described* by data on the Web.

5.2.1   Linked Data Principles

In the introduction we stated that Linked Data is a set of four best practices for publishing structured data on the Web [26]. Below, we state these four principles. The technology stack used for implementing these principles and publishing the data is described next, followed by the impact of the Linking Open Data (LOD) project [11], a direct consequence of the widespread adoption of these four principles.

1. **Use Uniform Resource Identifiers (URIs) as names for things.** Even though we primarily think of them as 'web addresses', URIs[1] are much more than just Uniform Resource Locators (URLs). In the broadest sense, a URI imposes constraints on, and sets a *standard* for [17], naming entities and units of data that people want to publish on the Web. In the case of HTML webpages, a URL serves nicely as the 'name' of the page. By using similar, albeit broader, standards for naming things, the first principle essentially ensures that we do not invent a new naming system from scratch. There are other benefits associated as well, as the second and third principles illustrate.
2. **Use Hypertext Transfer Protocol (HTTP) URIs so that the names can be looked up.** By associating HTTP lookup with URIs, the second principle ensures that the name of a thing is *dereferencable*. One simple way to do so is to ensure that URIs are also URLs. However, recall that the goal of Linked Data is to describe actual things, not just the description of things. By using techniques such as re-direction in conjunction with the first and second principles, it is possible to maintain this distinction. Intuitively, one could use a URI (not a URL) to name the thing itself, but when dereferenced, a re-direction could be used to direct the user to a URL which describes the thing. This is an elegant, rigorous way of ensuring that the names of things, as well as the descriptions of those things, could co-exist as separate artifacts on the Web.
3. **When a URI is looked up, provide useful information by using established standards such as RDF and SPARQL to publish and access information.** Resource Description Framework (RDF) and the SPARQL query language are important standards that have been developed over more than a decade by long-time researchers in the Semantic Web and Description Logics communities. The third principle ensures that when a URI is looked up, the data is not delivered in some ad-hoc format (e.g., as an Excel file), but instead conforms to well-established, open standards that can be consumed in a predictable way by a machine. Because the first and second principles ensure the use of HTTP and URIs, it is easier than it would be otherwise to implement the third principle. This also illustrates that the rules are not necessarily independent but build upon each other for effectiveness.

---

[1] In actuality, the first principle if even broader, allowing the use of *internationalized* resource identifiers rather than just URIs for naming things.

4. **Include links to other URIs, so that more relevant things can be discovered through navigation.** In a previous chapter, we covered the problem of Entity Resolution, which was a step designed to ensure that two or more entity 'mentions' referring to the same underlying entity would get 'resolved'. The mechanics on how such a resolution would happen, once the co-referent entities have been identified, were not described, since there is no one best practice. In the Linked Data scenario, a practitioner could simply publish an additional triple linking an entity in their dataset to equivalent entities in other datasets already existing as Linked Data. For example, as we cover later, DBpedia has emerged as a nexus for the openly published Linked Data on the Web, and since most entities in Wikipedia are included in DBpedia [8], linking entities in a dataset to DBpedia can often productively fulfill the fourth Linked Data principle. However, we also note that, while ER can be an important and well-defined mechanism for establishing links between entities in two different datasets, other relations can also be used. The knowledge graph embedding (KGE) techniques that we covered in the previous chapter could be a potent tool in this direction.

5.2.2 Technology Stack

The principles above illustrate that Linked Data is dependent on two technologies fundamental to the Web itself [27]: Uniform Resource Identifiers (URIs) and the Hypertext Transfer Protocol (HTTP). As we described earlier in the context of the first Linked Data principle, while Uniform Resource *Locators* (URLs) have become familiar as addresses for documents that can be located on the Web, Uniform Resource *Identifiers* provide a more generic means to identify any entity that exists in the world.

In the context of the second principle, where entities are identified by URIs using schemes such as http:// and https://, they can be looked up by dereferencing the URI leveraging the HTTP protocol. Thus, the HTTP protocol provides a simple, yet universal, mechanism either for retrieving resources that *can* be serialized as bytes, or retrieving (e.g., by using re-direction) *descriptions* of entities that cannot *physically* be uploaded and sent across networks.[2]

URIs and HTTP are supplemented by the RDF model, which is critical to implementing the vision of the Semantic Web and Linked Data. The use of RDF, and other technologies like SPARQL that execute on top of RDF triplestores to enable access to the data, is in response to the third Linked Data principle which requires information retrieved to be useful (importantly, both to humans and machines).

---

[2]For example, one could use the protocol for retrieving the description of a book, since the protocol cannot be used for sending the book itself across a network.

Although HTML allows us to structure and link documents on the Web, RDF provides a generic, graph-based data model with which to structure and link data that describes things (i.e. entities) in the world and the varied properties (typed links) that exist between entities. These typed links can have pre-defined semantics, such as owl:sameAs, since they come from a standard (widely used) upper-level vocabulary like SKOS, Dublin Core or RDFS [5]. These higher-level vocabularies are especially useful in facilitating the re-use of ontological terms and properties, ensuring more homogeneity than might be found. For example, properties like owl:sameAs are overwhelmingly used to capture, and publish, the results of Entity Resolution [96] and fulfill requirements such as the fourth Linked Data principle.

5.2.3   Linking Open Data

Because the Linked Data principles are recommended best practices, their success can only be measured in terms of impact and adoption. Perhaps the most visible evidence of impact has been an on-going, decentralized and international effort called the Linking Open Data (LOD) project (Fig. 5.1),[3] which has been described as 'a grassroots community effort founded in January 2007 and supported by the W3C Semantic Web Education and Outreach Group[4]'. The main goal of the effort is to *bootstrap* the Web of Data, and the adoption of the Linked Data principles, by identifying existing, open-license datasets, converting these datasets to RDF in accordance with Linked Data principles, and publishing them on the Web. An auxiliary goal is to facilitate more publishing of such datasets, with the hope that they become discoverable and usable by virtue of following the principles (especially the fourth principle, which encourages inter-linking).

Historically, the earliest participants (still accounting for a major portion of activity on LOD) were university academics, and small companies looking to gain a competitive advantage with high-risk technology. However, LOD has since become considerably more diverse, with significant current involvement from major players in media, government and tech such as the BBC, Thomson Reuters, New York Times and the Library of Congress. We posit that this growth is enabled by the open nature of the project, where anyone can participate simply by publishing a dataset according to the Linked Data principles and by interlinking it with existing datasets (a special case of the fourth principle). Although the growth is not as super-linear anymore as it was in the early stages of LOD, the ecosystem has remained popular. In the next section, we describe one of the success cases (DBpedia), which has found adopters across the spectrum in natural language processing, Semantic Web, and knowledge discovery.

---

[3]http://esw.w3.org/topic/SweoIG/TaskForces/CommunityProjects/LinkingOpenData
[4]http://www.w3.org/2001/sw/sweo/

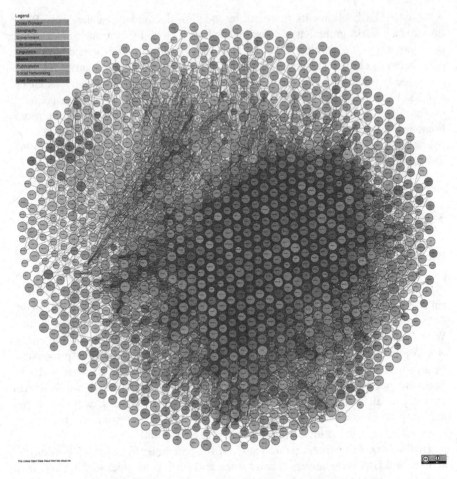

Fig. 5.1 A visualization of LOD datasets. Each node is a dataset and links represent connections between datasets in accordance with the fourth Linked Data principle. Colors represent domains e.g., social networking datasets are in grey. (Image courtesy of lod-cloud.net)

5.2.4   Example: DBpedia

The best example, and most well-known outcome of LOD, has been the DBpedia knowledge graph [8]. DBpedia was an early effort that sought to leverage the structured information on Wikipedia, which is itself a community-powered and crowdsourced encyclopedia. Figure 5.2 provides some intuition on how DBpedia would represent the entity 'Bob Marley' by extracting information from Bob Marley's infobox on his Wikipedia page, ontologizing it with respect to the DBpedia ontology, and rendering it as RDF. All four Linked Data principles are obeyed in this transformation process. DBpedia is available both as RDF dumps, and as a queryable SPARQL endpoint.

Bob Marley

From Wikipedia, the free encyclopedia

"Marley" redirects here. For other uses, see Bob Marley (disambiguation).

> This article **needs additional citations for verification**. Please help improve this article by adding citations to reliable sources. Unsourced material may be challenged and removed. *(March 2018) (Learn how and when to remove this template message)*

Robert Nesta Marley, OM (6 February 1945 – 11 May 1981) was a Jamaican singer-songwriter who became an international musical and cultural icon,[1][2][3][4][5] blending mostly reggae, ska, and rocksteady in his compositions. Starting out in 1963 with the group the Wailers, he forged a distinctive songwriting and vocal style that would later resonate with audiences worldwide. The Wailers would go on to release some of the earliest reggae records with producer Lee "Scratch" Perry.[6]

After the Wailers disbanded in 1974,[7] Marley pursued a solo career upon his relocation to England that culminated in the release of the album *Exodus* in 1977, which established his worldwide reputation and elevated his status as one of the world's best-selling artists of all time, with sales of more than 75 million records.[8][9] *Exodus* stayed on the British album charts for 56 consecutive weeks. It included four UK hit singles: "Exodus", "Waiting in Vain", "Jamming", and "One Love". In 1978, he released the album *Kaya*, which included the hit singles "Is This Love" and "Satisfy My Soul". The greatest hits album, *Legend*, was released in 1984, three years after Marley died. It subsequently became the best-selling reggae album of all time.

Diagnosed with acral lentiginous melanoma in 1977, Marley died on 11 May 1981 in Miami at age 36. He was a committed Rastafari who infused his music with a sense of spirituality.[10][242][11] He is credited with popularising reggae music around the world and served as a symbol of Jamaican culture and identity. Marley has also evolved into a global symbol and has inspired a significant merchandise industry.

Contents [hide]

1 Early life and career
2 Musical career
  2.1 1962–72: Early years
  2.2 1972–74: Move to Island Records
  2.3 1974–76: Line-up changes and shooting
  2.4 1976–79: Relocation to England
  2.5 1979–81: Later years
3 Illness and death
4 Legacy
  4.1 Awards and honours
  4.2 Other tributes
5 Personal life
  5.1 Religion
  5.2 Family
  5.3 Association football
  5.4 Personal views
    5.4.1 Pan-Africanism
    5.4.2 Cannabis
6 Discography
  6.1 Studio albums
  6.2 Live albums
7 See also
8 References
  8.1 Citations
  8.2 Sources
9 Further reading

Extract from infobox+ontologize+publish in RDF

The Honourable
Bob Marley
OM

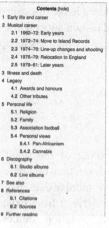

Marley performing in 1980

Born	Robert Nesta Marley, 6 February 1945, Nine Mile, Saint Ann Parish, British Jamaica
Died	11 May 1981 (aged 36), Miami, Florida, U.S.
Cause of death	Melanoma (skin cancer)
Other names	Donald Marley · Tuff Gong
Occupation	Singer · songwriter · musician
Home town	Trenchtown, Kingston, Jamaica
Spouse(s)	Alpharita Anderson Marley (m. 1966)
Children	Sharon Marley Prendergast (adopted), Cedella Marley, David Nesta "Ziggy" Marley, Stephen Robert Nesta Marley, Rohan Anthony Marley, Julian Ricardo Marley, Ky-Mani Marley, Damian Robert Nesta Marley

About: Bob Marley

An Entity in Data Space: ec2-174-129-126-64.compute-1.amazonaws.com

Robert "Bob" Nesta Marley OM (February 6, 1945 – May 11, 1981) was a Jamaican singer-songwriter and musician.

Property	Value
dbpedia-owl:associatedBand	• dbpedia:The_Upsetters • dbpedia:The_Wailers_%28reggae_band%29 • dbpedia:I_Threes • dbpedia:Wailers_Band
dbpedia-owl:associatedMusicalArtist	• dbpedia:The_Upsetters • dbpedia:The_Wailers_%28reggae_band%29 • dbpedia:I_Threes • dbpedia:Wailers_Band
dbpedia-owl:background	• solo_singer
dbpedia-owl:birthdate	• 1945-02-06 (xsd:date)
dbpedia-owl:birthname	• Robert Nesta Marley
dbpedia-owl:birthplace	• dbpedia:Saint_Ann_Parish%2C_Jamaica • dbpedia:Jamaica
dbpedia-owl:deathdate	• 1981-05-11 (xsd:date)
dbpedia-owl:deathplace	• dbpedia:Miami%2C_Florida • dbpedia:Florida

Fig. 5.2 An example DBpedia dashboard fragment describing Bob Marley

DBpedia generally occupies the central position in LOD diagrams because of two inter-related characteristics: (1) it is open-world and contains many entities, concepts and properties of interest, since it is ultimately described from Wikipedia infoboxes, and (2) for various reasons, many publishers on LOD have chosen to link to DBpedia entities to fulfill the fourth Linked Data principle. Although DBpedia is not dynamically fresh in the same vein as Wikipedia, which is constantly maintained by a community-enforced system of edits, revisions and additions, DBpedia is *periodically* derived from Wikipedia by executing extractors on Wikipedia dumps. Thus it is *relatively* fresh compared to more static datasets on LOD.

Overall, DBpedia continues to be well-maintained and widely used. In part, this is because of its dependence on Wikipedia, which has continued to be popular, but also because numerous applications across the Semantic Web, knowledge discovery and NLP communities now leverage it for weak supervision and distant supervision-style problems. Just like Wikipedia, DBpedia is also multi-lingual, which opens up even more applications.

5.3 Google Knowledge Vault

The Google Knowledge Vault, which indirectly populates some of the search features in Google, is a Web-scale probabilistic knowledge base that combines *extractions* from Web content (obtained via analytics over text, tabular data, page structure, and even human annotations) with *prior knowledge* derived from existing knowledge repositories [55]. Because these are distinct information sources, supervised machine learning methods have to be used for knowledge *fusion*. At the time of publication, this Knowledge Vault (KV) was assumed to be substantially bigger than any previously published structured knowledge repository, and featured a probabilistic inference system that could compute calibrated probabilities of assertion correctness. The authors of the Knowledge Vault paper report results from several studies and experiments illustrating the utility of the method [55].

Fundamentally, the KV was no different at an architectural level (see Fig. 5.3 for the architectural description of the KV) than the workflow proposed in this book. That is, there were three main components. The first component was a layer of extractors. Recall that, in an earlier chapter, we provided extensive details on information extraction, which is among the first steps in constructing a domain-specific knowledge graph from raw data. The KV is no different, although it is not single-domain. Extraction methods include text (including relation extraction, although the authors run standard methods at much larger scale), and Web IE methods like parsing the DOM trees of HTML pages, and also tables extracted from HTML. The KV also contains data from pages annotated manually with elements from ontologies like schema.org and openGraphProtocol.org [67, 69]. Schema.org is described in more detail in the following section.

However, rather than learn all its knowledge about the world from just IE, the KV also relied on prior knowledge by using graph-based priors. In essence,

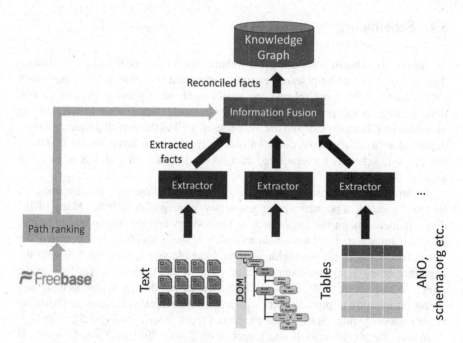

Fig. 5.3 A schematic showing the construction and workflow of the Google Knowledge Vault

these systems would learn the prior probability of each possible triple, based on triples stored in an existing KB. Technically, the procedure was different from the KGEs that we covered in the previous chapter; however, the concept was similar. For example, one of the applications of knowledge graph embeddings was *triples classification*, namely, the task of determining the probability of correctness of a hypothetical triple, given all the triples observed during training. Incorporating graph-based priors into the KV relied on a similar intuition.

Finally, the third key innovation in the KV architecture was an information or knowledge fusion box that would take the outputs of extractors, as well as those based on graph-based priors, and reconcile the facts based on supervised machine learning. Knowledge fusion was like the test phase in a triples classification system. In the actual paper, the authors consider several principled supervised machine learning methods.

Although it is not known whether the KV constitutes the core technology powering the current iteration of the Google Knowledge Graph [164], its influence on the construction of Web-scale knowledge graphs from heterogeneous structured and unstructured data sources is undeniable. The effort has proven difficult to replicate in non-industrial settings, however. Some of the components, like information fusion, have also been superseded by recent innovations such as knowledge graph embeddings. However, the role of extractors and the leverage of prior knowledge in reconciling contradictions continue to be important in existing KG construction pipelines.

5.4 Schema.org

Schema.org is a shared vocabulary that webmasters can use to structure metadata on their websites and to help search engines understand the content being published [69]. Although the term *schema.org* would seem to suggest a website (which it is, leading to the project's homepage) it is contextually used to refer to the vocabulary itself, the markup on the webpages as well as the overall project, which is described as 'a collaborative, community activity with a mission to create, maintain, and promote schemas for structured data on the Internet, on web pages, in email messages, and beyond'.

As an example, consider the movie 'Bohemian Rhapsody' as described by Rotten Tomatoes, a popular movie and review aggregation website. Many of the dynamic elements on the page (such as the reviews and the Tomatometer rating) have semantics associated with them according to the concepts and properties in the schema.org vocabulary. We highlight some example snippets in Fig. 5.4, with the aggregate rating being one example of an element that is visually rendered on the screen. When a search engine like Google scrapes this data, it is able to make use of this information to provide users with a better search experience (e.g., providing better answers to queries like 'rotten tomatoes top movies', see Fig. 5.5).

In fact, the initiative itself was launched on 2 June 2011 by Bing, Google and Yahoo! to create, support and develop a common set of schemas for structured

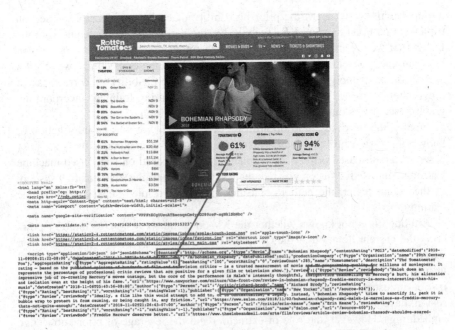

Fig. 5.4 Example of schema.org snippets embedded in Rotten Tomato's HTML webpages

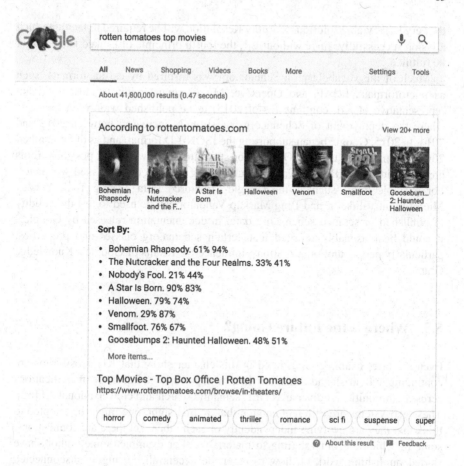

Fig. 5.5 Example of dynamic, enhanced search by commercial search providers using extracted schema.org information

markup on webpages. This would indicate that enhanced search has been one of the primary goals for the initiative. Later that year, Yandex joined the initiative. The main motivation behind using schema.org is that the markup can be recognized by search engine spiders and other parsers, which enables a layer of semantics to be incorporated into search engine optimization. Recall that the Semantic Web (which includes Linked Data [26]) vision was similar [18], but the principles of Linked Data do not necessarily apply to schema.org. As one example, much of the published schema.org markup does not even attempt to link to other schema.org markup. Consequently, the schema.org ecosystem is less like KGs in the LOD universe [11], and more like isolated knowledge fragments that provide local context to the webpages in which they are embedded. However, there have been efforts to try and integrate LOD and schema.org with some recent papers proposing to find and add schema.org fragments to the LOD. The work is still in its early stages, however. At the very least, viable solutions to Entity Resolution will be required to facilitate such a vision successfully, since without ER, the fourth principle of Linked Data cannot be fulfilled.

Much of the vocabulary on Schema.org was inspired by earlier formats, such as microformats, FOAF, and OpenCyc. Microformats, with its most dominant representative hCard, continue (as of 2015) to be published widely on the web, where the deployment of Schema.org has strongly increased between 2012 and 2014. In 2015, Google began supporting the JSON-LD format, and as of September, 2017 recommended using JSON-LD for structured data whenever possible. Tools are also widely available to validate schema.org markup on published webpages. For example, tools such as the Google Structured Data Testing Tool, Yandex Microformat validator, and Bing Markup Validator can be used to test the validity of published or scraped schema.org data. In documentation released by Google,[5] it could be reasonably inferred that certain schema.org classes and properties, particularly people and organizations, influence the results of Google's Knowledge Graph.

5.5   Where is the Future Going?

Even the brief examples illustrated in this chapter show that KG ecosystems are continuing to flourish and come into their own, powering a full range of applications across communities as diverse as NLP, semantic search and conversational AI [65]. These ecosystems are different enough that, at first sight, one might be tempted to think that they are evolving independently of each other. Yet there are connections, some of which are only starting to materialize. For example, some authors have started publishing work on how to reconcile decentralized, highly disconnected schema.org knowledge fragments with the larger Linked Data ecosystem. Freebase, which was used to power the KV and has since been taken over by Google, was an essential part of the Linking Open Data project in its initial phase, and has since been replaced with Wikidata [174]. Nevertheless, despite all these connections, the question remains: is there some way to reconcile all of these different KGs under a single umbrella, one that is open and accessible to all?

An example of one such initiative, still in a seedling stage, is the Open Knowledge Network[6] (OKN), which is attempting to jumpstart and realize the long-held vision of a *common semantic information infrastructure* for the future [116]. Recognizing the motivation that natural interfaces to large knowledge structures have the potential to impact science, education and business to an extent comparable to the WWW, the OKN initiative argues that KG-centric services like Alexa and

---

[5]https://developers.google.com/search/docs/guides/enhance-site#add-your-sites-name-logo-and-social-links

[6]https://okfn.org/network/

Cortana, or the Google search engine, are limited in their scope of knowledge, not open to direct access or contributors beyond their corporate firewalls, and can only answer relatively limited questions in their business areas. OKN wants to pioneer an architecture that will allow stakeholders to encode knowledge for their topics of interest and be able to hook them into the larger network, without having to go through gatekeepers (such as Google or Apple). Furthermore, once the knowledge is encoded, access to this should not be restricted to a small priesthood of SQL or other programmatic interface users. There will be a wide range of interfaces, including natural language interfaces, graphical interfaces and visualizations which no one has even invented yet. Developers will be able to independently create more sophisticated programs for answering queries, providing summaries that help regular people make decisions in their lives.

In order to realize the vision of an open Web-scale knowledge network, an attempt like the Google Knowledge Vault is required but at a scale that (arguably) is even more extensive. As ambitious as this may sound, the steering members of OKN argue that the technologies for realizing such a network already exist. However, it is also undeniable that there are many hurdles in realizing such an ambition, including obvious issues of cost, incentives and maintenance. From a purely research standpoint however, the OKN would be far more comprehensive than any existing KG ecosystem, and would likely trigger revolutionary advances in KG-centric applications.

While the OKN is likely a longer-term initiative that will require the coalescing of multiple research communities, there are several medium-term research challenges that researchers have already started focusing on. Entity Resolution continues to be a vital area of research, especially considering our arguments in the earlier chapter on how existing systems continue to fall short on several requirements that are essential for conducting ER on large KGs. Information extraction (IE) also continues to advance each year, though some kinds of IE are witnessing more attention than others. NER research, for example, seems to have plateaued, but relation and event extraction systems continue to be presented, even at the time of writing. Low-supervision IE has also seen a surge of interest. Finally, IE for languages other than English, and particularly for 'low-resource' languages for which good translation services are not available, has seen an increased surge in research interest due to programs funded by agencies like DARPA.

Glossary

Knowledge Graph A knowledge graph (KG) is a directed, labeled multi-relational graph where nodes typically represent either entities or the attributes of entities, and (labeled) edges represent either relationships between entity-entity pairs or properties of entities. The simplest way to serialize a KG is as a set of triples, where each triple is of the form (h, r, t) and represents an edge in the graph.

Ontology Although open-world KGs like Wikidata and Freebase have existed for a while, domain-specific KGs often derive their semantics and constraints from an underlying ontology. Although a deep definition of an ontology is not within the scope of this book, a simple definition is that an ontology is a set of terms defining the domain of interest. In the KG community, this amounts to a graph-like structure that contains concepts and relationships. A special relationship, is-a, serves as the glue relationship between the KG and the ontology. Evaluating when one ontology is 'better' than another continues to be a hotly debated issue, since it is not clear how to measure the goodness of an ontology using purely objective metrics.

Information Extraction When constructing domain-specific KGs, information extraction (IE) is the first set of algorithms that must be applied. IE refers to a set of techniques for ingesting natural language or HTML (and also other heterogeneous data that are neither structured nor natural language) and extracting useful information from them, usually with reference to an underlying ontology. IE today is often broken up into at least three sub-problems, each of which is important enough and challenging enough in its own right: Named Entity Recognition, Relation Extraction and Event Extraction.

Named Entity Recognition Named Entity Recognition (NER) is the best known sub-problem in Information Extraction. Although 'anything' in principle could be a named entity, in practice, named entities constitute instances of ontological types like persons, locations, organizations, facilities etc. In domain-specific applications, named entities can be esoteric and highly dependent on the domain e.g., physical attributes may be more common in the e-commerce domain than in

M. Kejriwal, *Domain-Specific Knowledge Graph Construction*, SpringerBriefs in Computer Science, https://doi.org/10.1007/978-3-030-12375-8

domains like politics, academia or medicine. Pre-trained NER systems are useful for extracting generic types, but IE techniques have to be applied to the domain-specific cases. Supervised, semi-supervised and unsupervised methods for IE currently exist, and more recently, deep learning and representation learning have become very popular for achieving state-of-the-art performance.

Word Embedding Feature engineering has always been a bottleneck in traditional machine learning pipelines, especially for natural language processing (NLP) applications like NER. In recent times however, word embedding models have emerged as an efficient and powerful means of vectorizing words, documents and even graphs into low-dimensional, continuous spaces. These vectors, when optimized using a relatively simple notion of context, yield remarkable insights in the vector space, such as analogies and semantic clustering. Multiple word embedding algorithms now exist, although the original innovations are still widely used. Word embeddings have generally been adopted in favor of heavily engineered feature pipelines across multiple application-oriented communities in machine learning and knowledge discovery.

Relation Extraction After NER, Relation Extraction (RE) is the next most important step in an IE pipeline, and is essential for knowledge graph construction. It is uncommon to extract n-ary relations with n greater than 2; even for binary relations, performance is relatively poor compared to NER. Relation Extraction can be framed as a classification problem assuming the entities have been correctly extracted. Other ways of framing the problem also exist. Similar to NER, deep learning has emerged as an important technique for tackling RE.

Event Extraction Event extraction is yet another IE sub-problem, but one that tends to be limited to certain domains and ontologies. Events are typically identified by 'triggers' e.g., the word 'hit' might trigger an 'attack' event type, and tend to involve multiple arguments and relations. It is not unreasonable to think of an event as a 'second-order' entity for that reason. It is generally believed that much more research is required on event extraction before performance will reach acceptable levels for broader consumption. Most event extraction papers still tend to focus on ontologies like ACE and CAMEO, which are broad but, by no means, complete. It is unknown whether any of the current techniques, including the state-of-the-art, would be able to adapt with relatively low overhead if a new (domain) ontology were to be introduced.

Entity Resolution Entity Resolution (ER) is the problem of algorithmically determining when two instances ('entities') in the KG are the same underlying entity. The problem has been around for more than 50 years, with patient linking and census being the earliest applications. ER has been studied under many guises, including record linkage, instance matching and deduplication. Just like many of the other techniques in this book, supervised, semi-supervised and unsupervised solutions exist. Performance of ER systems can vary widely depending both on the training regime, the amount of data available and the domain. On some domains, such as geopolitical events, ER continues to suffer from performance issues compared to more traditional domains like census and publications.

Blocking   Blocking refers to an important class of algorithmic techniques that are almost always included as the first step in a typical two-step ER workflow. Blocking is defined (in the most general case) as inexpensive clustering of approximately similar entities. By doing such clustering in sub-quadratic time, exhaustive pairwise comparisons can be avoided, leading to significant savings even for moderately sized datasets. Blocking has been explored for many decades now, and more recently, the automatic learning of blocking keys has become an important topic of research. Blocking for KGs is not as well-studied as blocking for relational databases.

Knowledge Graph Embedding   Similar to a word embedding, a knowledge graph embedding (KGE) can be used to embed entities and relationships in a KG into a low-dimensional continuous space. Most successful KGE models are translational models, such as TransE and TransD, and rely on the same kinds of analogical intuitions as traditional word embedding algorithms. Recently, a lot of research has been going into how to incorporate 'extra' information into the KGE models, be it external text corpora, ontologies, rules, temporal information etc. In general, there are empirical benefits to including more information into the KGE model, but one has to be careful about diluting the domain-specific value of the KG itself when bringing in generic external sources.

Linked Data   Linked Data is a set of four principles that guide the publication of (structured) data on the Web. Linked Data has continued to become popular, leading to the Linking Open Data (LOD) project, which contains datasets published openly on the Web using Linked Data standards. LOD now includes hundreds of datasets, including DBpedia, which is derived from Wikipedia, as its central hub. It is also the backbone for the broader Semantic Web ecosystem. Several rich applications are powered by LOD.

Knowledge Graph Ecosystem   In the broadest sense, a knowledge graph ecosystem is a community, rather than a collection of datasets. Like any community, such an ecosystem is guided by its own social norms and incentive structures. As the name suggests, a KG ecosystem is centered around using KGs as a prominent technology, but the definition of a KG, and even a domain, will differ slightly based on the ecosystem. Norms can be radically different, even for Web-based ecosystems. For example, schema.org, launched by search engines in the earlier part of this decade, encourages isolated knowledge fragments that can be embedded in HTML and easily found by search engines, while Linked Data emphasizes connectivity and is agnostic to search as a specific application. Lately, there have been efforts to map and possibly reconcile such ecosystems. Whether this is really possible will depend on both social and technological factors.

References

1. Achichi, M., Cheatham, M., Dragisic, Z., Euzenat, J., Faria, D., Ferrara, A., Flouris, G., Fundulaki, I., Harrow, I., Ivanova, V., et al.: Results of the ontology alignment evaluation initiative 2016. In: OM: Ontology Matching, pp. 73–129. No commercial editor. (2016)
2. Agichtein, E., Gravano, L.: Snowball: extracting relations from large plain-text collections. In: Proceedings of the Fifth ACM Conference on Digital Libraries, pp. 85–94. ACM (2000)
3. Ahn, D.: The stages of event extraction. In: Proceedings of the Workshop on Annotating and Reasoning About Time and Events, pp. 1–8. Association for Computational Linguistics (2006)
4. Alfonseca, E., Manandhar, S.: An unsupervised method for general named entity recognition and automated concept discovery. In: Proceedings of the 1st International Conference on General WordNet, Mysore, pp. 34–43 (2002)
5. Allemang, D., Hendler, J.: Semantic Web for the Working Ontologist: Effective Modeling in RDFS and OWL. Elsevier, Amsterdam (2011)
6. Arasu, A., Garcia-Molina, H.: Extracting structured data from web pages. In: Proceedings of the 2003 ACM SIGMOD International Conference on Management of Data, pp. 337–348. ACM (2003)
7. Ashburner, M., Ball, C.A., Blake, J.A., Botstein, D., Butler, H., Cherry, J.M., Davis, A.P., Dolinski, K., Dwight, S.S., Eppig, J.T., et al.: Gene ontology: tool for the unification of biology. Nat. Genet. **25**(1), 25 (2000)
8. Auer, S., Bizer, C., Kobilarov, G., Lehmann, J., Cyganiak, R., Ives, Z.: Dbpedia: a nucleus for a web of open data. In: The Semantic Web, pp. 722–735. Springer, Berlin (2007)
9. Bach, N., Badaskar, S.: A survey on relation extraction. Language Technologies Institute, Carnegie Mellon University (2007)
10. Banko, M., Cafarella, M.J., Soderland, S., Broadhead, M., Etzioni, O.: Open information extraction from the web. In: IJCAI, vol. 7, pp. 2670–2676 (2007)
11. Bauer, F., Kaltenböck, M.: Linked Open Data: The Essentials. Edition mono/monochrom, Vienna (2011)
12. Baxter, R., Christen, P., Churches, T., et al.: A comparison of fast blocking methods for record linkage. In: ACM SIGKDD, vol. 3, pp. 25–27. Citeseer (2003)
13. Benajiba, Y., Diab, M., Rosso, P.: Arabic named entity recognition using optimized feature sets. In: Proceedings of the Conference on Empirical Methods in Natural Language Processing, pp. 284–293. Association for Computational Linguistics (2008)
14. Bengio, Y., Courville, A., Vincent, P.: Representation learning: a review and new perspectives. IEEE Trans. Pattern Anal. Mach. Intell. **35**(8), 1798–1828 (2013)

© The Author(s), under exclusive license to Springer Nature Switzerland AG 2019
M. Kejriwal, *Domain-Specific Knowledge Graph Construction*, SpringerBriefs in Computer Science, https://doi.org/10.1007/978-3-030-12375-8

15. Benjelloun, O., Garcia-Molina, H., Menestrina, D., Su, Q., Whang, S.E., Widom, J.: Swoosh: a generic approach to entity resolution. VLDB J.: Int. J. Very Large Data Bases **18**(1), 255–276 (2009)

16. Berant, J., Srikumar, V., Chen, P.C., Vander Linden, A., Harding, B., Huang, B., Clark, P., Manning, C.D.: Modeling biological processes for reading comprehension. In: Proceedings of the 2014 Conference on Empirical Methods in Natural Language Processing (EMNLP), pp. 1499–1510 (2014)

17. Berners-Lee, T., Fielding, R., Masinter, L.: Uniform resource identifier (URI): generic syntax. Technical report (2004)

18. Berners-Lee, T., Hendler, J., Lassila, O.: The semantic web. Sci. Am. **284**(5), 34–43 (2001)

19. Bhaskaran, S., Rafeeque, P.: A survey on relation extraction methodologies from unstructured text. In: Emerging Trends in Engineering, Science and Technology for Society, Energy and Environment, pp. 869–874. CRC Press, Leiden (2018)

20. Bhattacharya, I., Getoor, L.: A latent dirichlet model for unsupervised entity resolution. In: Proceedings of the 2006 SIAM International Conference on Data Mining, pp. 47–58. SIAM (2006)

21. Bhattacharya, I., Getoor, L.: Collective entity resolution in relational data. ACM Trans. Knowl. Discov. Data (TKDD) **1**(1), 5 (2007)

22. Bick, E.: A named entity recognizer for Danish. In: LREC. Citeseer (2004)

23. Bilenko, M., Mooney, R.J.: Adaptive duplicate detection using learnable string similarity measures. In: Proceedings of the Ninth ACM SIGKDD International Conference on Knowledge Discovery and Data Mining, pp. 39–48. ACM (2003)

24. Bilke, A., Naumann, F.: Schema matching using duplicates. In: 21st International Conference on Data Engineering, 2005 (ICDE 2005). Proceedings, pp. 69–80. IEEE (2005)

25. Bizer, C.: The emerging web of linked data. IEEE Intell. Syst. **24**(5), 87–92 (2009)

26. Bizer, C., Heath, T., Berners-Lee, T.: Linked data: the story so far. In: Semantic Services, Interoperability and Web Applications: Emerging Concepts, pp. 205–227. IGI Global (2011)

27. Bizer, C., Heath, T., Idehen, K., Berners-Lee, T.: Linked data on the web (ldow2008). In: Proceedings of the 17th International Conference on World Wide Web, pp. 1265–1266. ACM (2008)

28. Björne, J., Heimonen, J., Ginter, F., Airola, A., Pahikkala, T., Salakoski, T.: Extracting complex biological events with rich graph-based feature sets. In: Proceedings of the Workshop on Current Trends in Biomedical Natural Language Processing: Shared Task, pp. 10–18. Association for Computational Linguistics (2009)

29. Blei, D.M., Ng, A.Y., Jordan, M.I.: Latent dirichlet allocation. J. Mach. Learn. Res. **3**, 993–1022 (2003)

30. Bordes, A., Usunier, N., Garcia-Duran, A., Weston, J., Yakhnenko, O.: Translating embeddings for modeling multi-relational data. In: Advances in Neural Information Processing Systems, pp. 2787–2795 (2013)

31. Bordes, A., Weston, J., Collobert, R., Bengio, Y., et al.: Learning structured embeddings of knowledge bases. In: AAAI, vol. 6, p. 6 (2011)

32. Brin, S.: Extracting patterns and relations from the world wide web. In: International Workshop on the World Wide Web and Databases, pp. 172–183. Springer (1998)

33. Cao, Y., Chen, Z., Zhu, J., Yue, P., Lin, C.Y., Yu, Y.: Leveraging unlabeled data to scale blocking for record linkage. In: IJCAI Proceedings-International Joint Conference on Artificial Intelligence, vol. 22, p. 2211 (2011)

34. Chakrabarti, K., Chaudhuri, S., Cheng, T., Xin, D.: A framework for robust discovery of entity synonyms. In: Proceedings of the 18th ACM SIGKDD International Conference on Knowledge Discovery and Data Mining, pp. 1384–1392. ACM (2012)

35. Chambers, N., Jurafsky, D.: Unsupervised learning of narrative event chains. In: Proceedings of ACL-08: HLT, pp. 789–797 (2008)

36. Chang, C.H., Kayed, M., Girgis, M.R., Shaalan, K.F.: A survey of web information extraction systems. IEEE Trans. Knowl. Data Eng. **18**(10), 1411–1428 (2006)

37. Chang, C.H., Kuo, S.C.: Olera: semisupervised web-data extraction with visual support. IEEE Intell. Syst. **19**(6), 56–64 (2004)
38. Chang, C.H., Lui, S.C.: Iepad: information extraction based on pattern discovery. In: Proceedings of the 10th International Conference on World Wide Web, pp. 681–688. ACM (2001)
39. Christen, P.: Automatic record linkage using seeded nearest neighbour and support vector machine classification. In: Proceedings of the 14th ACM SIGKDD International Conference on Knowledge Discovery and Data Mining, pp. 151–159. ACM (2008)
40. Christen, P.: Febrl-: an open source data cleaning, deduplication and record linkage system with a graphical user interface. In: Proceedings of the 14th ACM SIGKDD International Conference on Knowledge Discovery and Data Mining, pp. 1065–1068. ACM (2008)
41. Christen, P.: Data Matching: Concepts and Techniques for Record Linkage, Entity Resolution, and Duplicate Detection. Springer, Heidelberg (2012)
42. Christen, P.: A survey of indexing techniques for scalable record linkage and deduplication. IEEE Trans. Knowl. Data Eng. **24**(9), 1537–1555 (2012)
43. Coates-Stephens, S.: The analysis and acquisition of proper names for the understanding of free text. Comput. Humanit. **26**(5–6), 441–456 (1992)
44. Cohen, W.W., Sarawagi, S.: Exploiting dictionaries in named entity extraction: combining semi-Markov extraction processes and data integration methods. In: Proceedings of the Tenth ACM SIGKDD International Conference on Knowledge Discovery and Data Mining, pp. 89–98. ACM (2004)
45. Collins, M.: Ranking algorithms for named-entity extraction: boosting and the voted perceptron. In: Proceedings of the 40th Annual Meeting on Association for Computational Linguistics, pp. 489–496. Association for Computational Linguistics (2002)
46. Collins, M., Singer, Y.: Unsupervised models for named entity classification. In: 1999 Joint SIGDAT Conference on Empirical Methods in Natural Language Processing and Very Large Corpora (1999)
47. Collobert, R., Weston, J.: A unified architecture for natural language processing: deep neural networks with multitask learning. In: Proceedings of the 25th International Conference on Machine Learning, pp. 160–167. ACM (2008)
48. Cowie, J., Lehnert, W.: Information extraction. Commun. ACM **39**(1), 80–91 (1996)
49. Crescenzi, V., Mecca, G., Merialdo, P., et al.: Roadrunner: towards automatic data extraction from large web sites. In: VLDB, vol. 1, pp. 109–118 (2001)
50. Cunningham, H.: Information extraction, automatic. In: Brown, K. (eds.) Encyclopedia of Language & Linguistics, vol. 5, 2nd edn., pp. 665–677. Elsevier, Oxford (2006)
51. Dai, A.M., Olah, C., Le, Q.V.: Document embedding with paragraph vectors. arXiv preprint arXiv:1507.07998 (2015)
52. Datar, M., Immorlica, N., Indyk, P., Mirrokni, V.S.: Locality-sensitive hashing scheme based on p-stable distributions. In: Proceedings of the Twentieth Annual Symposium on Computational Geometry, pp. 253–262. ACM (2004)
53. Dempster, A.P., Laird, N.M., Rubin, D.B.: Maximum likelihood from incomplete data via the em algorithm. J. R. Stat. Soc. Ser. B (Methodol.) **39**(1), 1–22 (1977)
54. Do, Q.X., Lu, W., Roth, D.: Joint inference for event timeline construction. In: Proceedings of the 2012 Joint Conference on Empirical Methods in Natural Language Processing and Computational Natural Language Learning, pp. 677–687. Association for Computational Linguistics (2012)
55. Dong, X., Gabrilovich, E., Heitz, G., Horn, W., Lao, N., Murphy, K., Strohmann, T., Sun, S., Zhang, W.: Knowledge vault: a web-scale approach to probabilistic knowledge fusion. In: Proceedings of the 20th ACM SIGKDD International Conference on Knowledge Discovery and Data Mining, pp. 601–610. ACM (2014)
56. Duan, S., Fokoue, A., Hassanzadeh, O., Kementsietsidis, A., Srinivas, K., Ward, M.J.: Instance-based matching of large ontologies using locality-sensitive hashing. In: International Semantic Web Conference, pp. 49–64. Springer (2012)

57. Elmagarmid, A.K., Ipeirotis, P.G., Verykios, V.S.: Duplicate record detection: a survey. IEEE Trans. Knowl. Data Eng. **19**(1), 1–16 (2007)
58. Esteban, C., Tresp, V., Yang, Y., Baier, S., Krompaß, D.: Predicting the co-evolution of event and knowledge graphs. In: 2016 19th International Conference on Information Fusion (FUSION), pp. 98–105. Ieee (2016)
59. Etzioni, O., Cafarella, M., Downey, D., Popescu, A.M., Shaked, T., Soderland, S., Weld, D.S., Yates, A.: Unsupervised named-entity extraction from the web: an experimental study. Artif. Intell. **165**(1), 91–134 (2005)
60. Euzenat, J., Shvaiko, P., et al.: Ontology Matching, vol. 18. Springer, Berlin (2007)
61. Fellegi, I.P., Sunter, A.B.: A theory for record linkage. J. Am. Stat. Assoc. **64**(328), 1183–1210 (1969)
62. Feng, J., Huang, M., Yang, Y., et al.: Gake: graph aware knowledge embedding. In: Proceedings of COLING 2016, the 26th International Conference on Computational Linguistics: Technical Papers, pp. 641–651 (2016)
63. Ferraram, A., Nikolov, A., Scharffe, F.: Data linking for the semantic web. Semant. Web: Ontol. Knowl. Base Enabled Tools Serv. Appl. **169**, 326 (2013)
64. Gaizauskas, R., Wakao, T., Humphreys, K., Cunningham, H., Wilks, Y.: University of sheffield: description of the lasie system as used for MUC-6. Technical report, Sheffield University (UK) (1995)
65. Gao, J., Galley, M., Li, L.: Neural approaches to conversational AI. In: The 41st International ACM SIGIR Conference on Research & Development in Information Retrieval, pp. 1371–1374. ACM (2018)
66. Getoor, L., Machanavajjhala, A.: Entity resolution: theory, practice & open challenges. Proc. VLDB Endowment **5**(12), 2018–2019 (2012)
67. Graham, W. (2012). Facebook developer tools. In: Beginning Facebook Game Apps Development, pp. 201–229. Apress
68. Grishman, R., Sundheim, B.: Message understanding conference-6: a brief history. In: COLING 1996: The 16th International Conference on Computational Linguistics, vol. 1 (1996)
69. Guha, R.V., Brickley, D., Macbeth, S.: Schema.org: evolution of structured data on the web. Commun. ACM **59**(2), 44–51 (2016)
70. Guo, S., Wang, Q., Wang, B., Wang, L., Guo, L.: Semantically smooth knowledge graph embedding. In: Proceedings of the 53rd Annual Meeting of the Association for Computational Linguistics and the 7th International Joint Conference on Natural Language Processing (Volume 1: Long Papers), vol. 1, pp. 84–94 (2015)
71. Guo, S., Wang, Q., Wang, L., Wang, B., Guo, L.: Jointly embedding knowledge graphs and logical rules. In: Proceedings of the 2016 Conference on Empirical Methods in Natural Language Processing, pp. 192–202 (2016)
72. Gusfield, D., Irving, R.W.: The Stable Marriage Problem: Structure and Algorithms. MIT Press, Cambridge (1989)
73. Hachey, B., Radford, W., Nothman, J., Honnibal, M., Curran, J.R.: Evaluating entity linking with wikipedia. Artif. Intell. **194**, 130–150 (2013)
74. Hearst, M.A.: Automatic acquisition of hyponyms from large text corpora. In: Proceedings of the 14th Conference on Computational Linguistics, vol. 2, pp. 539–545. Association for Computational Linguistics (1992)
75. Heath, T., Bizer, C.: Linked data: evolving the web into a global data space. Synth. Lect. Semant. Web: Theory Technol. **1**(1), 1–136 (2011)
76. Hernández, M.A., Stolfo, S.J.: Real-world data is dirty: data cleansing and the merge/purge problem. Data Min. Knowl. Discov. **2**(1), 9–37 (1998)
77. Hogue, A., Karger, D.: Thresher: automating the unwrapping of semantic content from the world wide web. In: Proceedings of the 14th International Conference on World Wide Web, pp. 86–95. ACM (2005)

78. Isozaki, H., Kazawa, H.: Efficient support vector classifiers for named entity recognition. In: Proceedings of the 19th International Conference on Computational Linguistics, vol. 1, pp. 1–7. Association for Computational Linguistics (2002)
79. Jain, A.K., Dubes, R.C.: Algorithms for Clustering Data. Prentice-Hall, Englewood Cliffs (1988)
80. Jansche, M.: Named entity extraction with conditional Markov models and classifiers. In: Proceedings of the 6th Conference on Natural Language Learning, vol. 20, pp. 1–4. Association for Computational Linguistics (2002)
81. Ji, H., Grishman, R.: Refining event extraction through cross-document inference. Proc. ACL-08: HLT, pp. 254–262 (2008)
82. Jiang, T., Liu, T., Ge, T., Sha, L., Li, S., Chang, B., Sui, Z.: Encoding temporal information for time-aware link prediction. In: Proceedings of the 2016 Conference on Empirical Methods in Natural Language Processing, pp. 2350–2354 (2016)
83. Jiang, X., Tresp, V., Huang, Y., Nickel, M.: Link prediction in multi-relational graphs using additive models. SeRSy **919**, 1–12 (2012)
84. Joulin, A., Grave, E., Bojanowski, P., Mikolov, T.: Bag of tricks for efficient text classification. arXiv preprint arXiv:1607.01759 (2016)
85. Kejriwal, M.: Populating entity name systems for big data integration. In: International Semantic Web Conference, pp. 521–528. Springer (2014)
86. Kejriwal, M.: Populating a Linked Data Entity Name System: A Big Data Solution to Unsupervised Instance Matching, vol. 27. IOS Press, Amsterdam (2016)
87. Kejriwal, M., Miranker, D.P.: An unsupervised algorithm for learning blocking schemes. In: 2013 IEEE 13th International Conference on Data Mining (ICDM), pp. 340–349. IEEE (2013)
88. Kejriwal, M., Miranker, D.P.: A two-step blocking scheme learner for scalable link discovery. In: Proceedings of the 9th International Conference on Ontology Matching, vol. 1317, pp. 49–60. CEUR-WS.org (2014)
89. Kejriwal, M., Miranker, D.P.: An unsupervised instance matcher for schema-free rdf data. Web Semant.: Sci. Serv. Agents World Wide Web **35**, 102–123 (2015)
90. Kejriwal, M., Szekely, P.: Neural embeddings for populated geonames locations. In: International Semantic Web Conference, pp. 139–146. Springer (2017)
91. Kim, J.D., Ohta, T., Pyysalo, S., Kano, Y., Tsujii, J.: Overview of bionlp'09 shared task on event extraction. In: Proceedings of the Workshop on Current Trends in Biomedical Natural Language Processing: Shared Task, pp. 1–9. Association for Computational Linguistics (2009)
92. Kimmig, A., Bach, S., Broecheler, M., Huang, B., Getoor, L.: A short introduction to probabilistic soft logic. In: Proceedings of the NIPS Workshop on Probabilistic Programming: Foundations and Applications, pp. 1–4 (2012)
93. Klement, E.P., Mesiar, R., Pap, E.: Triangular Norms, vol. 8. Springer, Cham (2013)
94. Kolb, L., Thor, A., Rahm, E.: Dedoop: efficient deduplication with hadoop. Proc. VLDB Endowment **5**(12), 1878–1881 (2012)
95. Kolb, L., Thor, A., Rahm, E.: Multi-pass sorted neighborhood blocking with mapreduce. Comput. Sci.-Res. Develop. **27**(1), 45–63 (2012)
96. Köpcke, H., Rahm, E.: Frameworks for entity matching: a comparison. Data Knowl. Eng. **69**(2), 197–210 (2010)
97. Kudo, T., Matsumoto, Y.: Chunking with support vector machines. In: Proceedings of the Second Meeting of the North American Chapter of the Association for Computational Linguistics on Language Technologies, pp. 1–8. Association for Computational Linguistics (2001)
98. Kumar, S.: A survey of deep learning methods for relation extraction. arXiv preprint arXiv:1705.03645 (2017)
99. Kushmerick, N., Weld, D.S., Doorenbos, R.: Wrapper Induction for Information Extraction, pp. 729–737. University of Washington, Washington (1997)

100. Lafferty, J., McCallum, A., Pereira, F.C.: Proceedings of the Eighteenth International Conference on Machine Learning (ICML '01), pp. 282–289. Morgan Kaufmann Publishers Inc., San Francisco (2001)

101. Leetaru, K., Schrodt, P.A.: Gdelt: global data on events, location, and tone, 1979–2012. In: ISA Annual Convention, vol. 2, pp. 1–49. Citeseer (2013)

102. Li, Q., Ji, H., Huang, L.: Joint event extraction via structured prediction with global features. In: Proceedings of the 51st Annual Meeting of the Association for Computational Linguistics (Volume 1: Long Papers), vol. 1, pp. 73–82 (2013)

103. Li, X., Morie, P., Roth, D.: Identification and tracing of ambiguous names: discriminative and generative approaches. In: Proceedings of the National Conference on Artificial Intelligence, 1999, pp. 419–424. AAAI Press/MIT Press, Menlo Park/Cambridge/London (2004)

104. Liao, S., Grishman, R.: Using document level cross-event inference to improve event extraction. In: Proceedings of the 48th Annual Meeting of the Association for Computational Linguistics, pp. 789–797. Association for Computational Linguistics (2010)

105. Lin, Y., Liu, Z., Sun, M.: Knowledge representation learning with entities, attributes and relations. Ethnicity 1, 41–52 (2016)

106. Lin, Y., Liu, Z., Sun, M., Liu, Y., Zhu, X.: Learning entity and relation embeddings for knowledge graph completion. In: AAAI, vol. 15, pp. 2181–2187 (2015)

107. Liu, X., Zhang, S., Wei, F., Zhou, M.: Recognizing named entities in tweets. In: Proceedings of the 49th Annual Meeting of the Association for Computational Linguistics: Human Language Technologies, vol. 1, pp. 359–367. Association for Computational Linguistics (2011)

108. Maaten, L.v.d., Hinton, G.: Visualizing data using t-SNE. J. Mach. Learn. Res. 9, 2579–2605 (2008)

109. Manning, C.D., Manning, C.D., Schütze, H.: Foundations of Statistical Natural Language Processing. MIT Press, Cambridge (1999)

110. Martínez, V., Berzal, F., Cubero, J.C.: A survey of link prediction in complex networks. ACM Comput. Surv. (CSUR) 49(4), 69 (2017)

111. McCallum, A.: Information extraction: distilling structured data from unstructured text. Queue 3(9), 4 (2005)

112. McCallum, A., Nigam, K., Ungar, L.H.: Efficient clustering of high-dimensional data sets with application to reference matching. In: Proceedings of the Sixth ACM SIGKDD International Conference on Knowledge Discovery and Data Mining, pp. 169–178. ACM (2000)

113. McClosky, D., Manning, C.D.: Learning constraints for consistent timeline extraction. In: Proceedings of the 2012 Joint Conference on Empirical Methods in Natural Language Processing and Computational Natural Language Learning, pp. 873–882. Association for Computational Linguistics (2012)

114. McClosky, D., Surdeanu, M., Manning, C.D.: Event extraction as dependency parsing. In: Proceedings of the 49th Annual Meeting of the Association for Computational Linguistics: Human Language Technologies, vol. 1, pp. 1626–1635. Association for Computational Linguistics (2011)

115. McDonald, D.: Internal and external evidence in the identification and semantic categorization of proper names. Proceedings of the 31st Annual Meeting of the Association for Computational Linguistics, 22–26 June 1993, Ohio State University, Columbus. ACL (1993)

116. Mena, E., Kashyap, V., Illarramendi, A., Sheth, A.: Domain specific ontologies for semantic information brokering on the global information infrastructure. In: Formal Ontology in Information Systems, vol. 46, pp. 269–283. IOS Press, Amsterdam (1998)

117. Mikheev, A., Moens, M., Grover, C.: Named entity recognition without gazetteers. In: Proceedings of the Ninth Conference on European Chapter of the Association for Computational Linguistics, pp. 1–8. Association for Computational Linguistics (1999)

118. Mikolov, T., Chen, K., Corrado, G., Dean, J., Sutskever, L., Zweig, G.: word2vec. https://code.google.com/p/word2vec (2013)

119. Mooney, R.J., Bunescu, R.: Mining knowledge from text using information extraction. ACM SIGKDD Explor. Newslett. **7**(1), 3–10 (2005)
120. Nadeau, D.: Semi-supervised named entity recognition: learning to recognize 100 entity types with little supervision. Ph.D. thesis, University of Ottawa (2007)
121. Nadeau, D., Sekine, S.: A survey of named entity recognition and classification. Lingvisticae Investigationes **30**(1), 3–26 (2007)
122. Ng, V.: Unsupervised models for conference resolution. In: Proceedings of the Conference on Empirical Methods in Natural Language Processing, pp. 640–649 (2008). Association for Computational Linguistics
123. Newcombe, H.B., Kennedy, J.M., Axford, S., James, A.P.: Automatic linkage of vital records. Science **130**(3381), 954–959 (1959)
124. Ngomo, A.C.N., Auer, S.: Limes-a time-efficient approach for large-scale link discovery on the web of data. In: IJCAI, pp. 2312–2317 (2011)
125. Ngomo, A.C.N., Lehmann, J., Auer, S., Höffner, K.: Raven-active learning of link specifications. In: Proceedings of the 6th International Conference on Ontology Matching, vol. 814, pp. 25–36. CEUR-WS.org (2011)
126. Ngomo, A.C.N., Lyko, K.: Eagle: efficient active learning of link specifications using genetic programming. In: Extended Semantic Web Conference, pp. 149–163. Springer (2012)
127. Ngomo, A.C.N., Lyko, K.: Unsupervised learning of link specifications: deterministic vs. non-deterministic. In: Proceedings of the 8th International Conference on Ontology Matching, vol. 1111, pp. 25–36. CEUR-WS.org (2013)
128. Ngomo, A.C.N., Lyko, K., Christen, V.: Coala–correlation-aware active learning of link specifications. In: Extended Semantic Web Conference, pp. 442–456. Springer (2013)
129. Nguyen, D.Q.: An overview of embedding models of entities and relationships for knowledge base completion. arXiv preprint arXiv:1703.08098 (2017)
130. Nguyen, T.H., Grishman, R.: Relation extraction: perspective from convolutional neural networks. In: Proceedings of the 1st Workshop on Vector Space Modeling for Natural Language Processing, pp. 39–48 (2015)
131. Nickel, M., Tresp, V., Kriegel, H.P.: Factorizing YAGO: scalable machine learning for linked data. In: Proceedings of the 21st International Conference on World Wide Web, pp. 271–280. ACM (2012)
132. Nie, F., Zhu, W., Li, X.: Unsupervised large graph embedding. In: AAAI, pp. 2422–2428 (2017)
133. Novak, B.: A survey of focused web crawling algorithms. In: Proceedings of SIKDD, pp. 55–58. Citeseer (2004)
134. Palmer, D.D., Day, D.S.: A statistical profile of the named entity task. In: Proceedings of the Fifth Conference on Applied Natural Language Processing, pp. 190–193. Association for Computational Linguistics (1997)
135. Pan, J.Z.: Resource description framework. In: Handbook on Ontologies, pp. 71–90. Springer, Berlin/Heidelberg (2009)
136. Pan, S.J., Yang, Q.: A survey on transfer learning. IEEE Trans. Knowl. Data Eng. **22**(10), 1345–1359 (2010)
137. Papadakis, G., Ioannou, E., Palpanas, T., Niederee, C., Nejdl, W.: A blocking framework for entity resolution in highly heterogeneous information spaces. IEEE Trans. Knowl. Data Eng. **25**(12), 2665–2682 (2013)
138. Patrick, J., Whitelaw, C., Munro, R.: Slinerc: the sydney language-independent named entity recogniser and classifier. In: Proceedings of the 6th Conference on Natural Language Learning, vol. 20, pp. 1–4. Association for Computational Linguistics (2002)
139. Paulheim, H.: Knowledge graph refinement: a survey of approaches and evaluation methods. Semant. Web **8**(3), 489–508 (2017)
140. Pennington, J., Socher, R., Manning, C.: Glove: global vectors for word representation. In: Proceedings of the 2014 Conference on Empirical Methods in Natural Language Processing (EMNLP), pp. 1532–1543 (2014)

141. Perozzi, B., Al-Rfou, R., Skiena, S.: Deepwalk: online learning of social representations. In: Proceedings of the 20th ACM SIGKDD International Conference on Knowledge Discovery and Data Mining, pp. 701–710. ACM (2014)

142. Piskorski, J., Tanev, H., Atkinson, M., Van Der Goot, E., Zavarella, V.: Online news event extraction for global crisis surveillance. In: Transactions on Computational Collective Intelligence V, pp. 182–212. Springer, Berlin/New York (2011)

143. Poibeau, T.: Dealing with metonymic readings of named entities. arXiv preprint cs/0607052 (2006)

144. Pujara, J., Augustine, E., Getoor, L.: Sparsity and noise: where knowledge graph embeddings fall short. In: Proceedings of the 2017 Conference on Empirical Methods in Natural Language Processing, pp. 1751–1756 (2017)

145. Pujara, J., Miao, H., Getoor, L., Cohen, W.: Knowledge graph identification. In: International Semantic Web Conference, pp. 542–557. Springer (2013)

146. Raghavan, H., Allan, J., McCallum, A.: An exploration of entity models, collective classification and relation description. In: KDD Workshop on Link Analysis and Group Detection, pp. 1–10 (2004)

147. Rahm, E., Bernstein, P.A.: A survey of approaches to automatic schema matching. VLDB J. **10**(4), 334–350 (2001)

148. Ramadan, B., Christen, P.: Unsupervised blocking key selection for real-time entity resolution. In: Pacific-Asia Conference on Knowledge Discovery and Data Mining, pp. 574–585. Springer (2015)

149. Ratinov, L., Roth, D.: Design challenges and misconceptions in named entity recognition. In: Proceedings of the Thirteenth Conference on Computational Natural Language Learning, pp. 147–155. Association for Computational Linguistics (2009)

150. Ravikumar, P., Cohen, W.W.: A hierarchical graphical model for record linkage. In: Proceedings of the 20th Conference on Uncertainty in Artificial Intelligence, pp. 454–461. AUAI Press (2004)

151. Ravin, Y., Leacock, C.: Polysemy: Theoretical and Computational Approaches. OUP, Oxford (2000)

152. Ravin, Y., Wacholder, N.: Extracting Names from Natural-Language Text. Citeseer, University Park (1997)

153. Raytheon, B.: Technologies. BBN accent event coding evaluation. Technical report (2015)

154. Riedel, S., McCallum, A.: Fast and robust joint models for biomedical event extraction. In: Proceedings of the Conference on Empirical Methods in Natural Language Processing, pp. 1–12. Association for Computational Linguistics (2011)

155. Riloff, E., Jones, R., et al.: Learning dictionaries for information extraction by multi-level bootstrapping. In: AAAI/IAAI, pp. 474–479 (1999)

156. Ritter, A., Clark, S., Etzioni, O., et al.: Named entity recognition in tweets: an experimental study. In: Proceedings of the Conference on Empirical Methods in Natural Language Processing, pp. 1524–1534. Association for Computational Linguistics (2011)

157. Rocktäschel, T., Singh, S., Riedel, S.: Injecting logical background knowledge into embeddings for relation extraction. In: Proceedings of the 2015 Conference of the North American Chapter of the Association for Computational Linguistics: Human Language Technologies, pp. 1119–1129 (2015)

158. Rocktäschel, T., Weidlich, M., Leser, U.: Chemspot: a hybrid system for chemical named entity recognition. Bioinformatics **28**(12), 1633–1640 (2012)

159. Rong, S., Niu, X., Xiang, E.W., Wang, H., Yang, Q., Yu, Y.: A machine learning approach for instance matching based on similarity metrics. In: International Semantic Web Conference, pp. 460–475. Springer (2012)

160. Salton, G., McGill, M.J.: Introduction to Modern Information Retrieval. McGraw-Hill, New York (1986)

161. Sarawagi, S., et al.: Information extraction. Found. Trends® Databases **1**(3), 261–377 (2008)

162. Settles, B.: Biomedical named entity recognition using conditional random fields and rich feature sets. In: Proceedings of the International Joint Workshop on Natural Language Processing in Biomedicine and Its Applications, pp. 104–107. Association for Computational Linguistics (2004)
163. Shadbolt, N., Berners-Lee, T., Hall, W.: The semantic web revisited. IEEE Intell. Syst. **21**(3), 96–101 (2006)
164. Singhal, A.: Introducing the knowledge graph: things, not strings. Off. Google Blog **5** (2012). https://www.blog.google/products/search/introducing-knowledge-graph-things-not/
165. Socher, R., Chen, D., Manning, C.D., Ng, A.: Reasoning with neural tensor networks for knowledge base completion. In: Advances in Neural Information Processing Systems, pp. 926–934 (2013)
166. Song, D., Heflin, J.: Automatically generating data linkages using a domain-independent candidate selection approach. In: The Semantic Web–ISWC 2011, pp. 649–664 (2011)
167. Suchanek, F.M., Kasneci, G., Weikum, G.: YAGO: a core of semantic knowledge. In: Proceedings of the 16th International Conference on World Wide Web, pp. 697–706. ACM (2007)
168. Tay, Y., Luu, A.T., Hui, S.C., Brauer, F.: Random semantic tensor ensemble for scalable knowledge graph link prediction. In: Proceedings of the Tenth ACM International Conference on Web Search and Data Mining, pp. 751–760. ACM (2017)
169. Thielen, C.: An approach to proper name tagging for German. arXiv preprint cmp-lg/9506024 (1995)
170. Tommasi, T., Patricia, N., Caputo, B., Tuytelaars, T.: A deeper look at dataset bias. In: Domain Adaptation in Computer Vision Applications, pp. 37–55. Springer, Cham (2017)
171. Tsuruoka, Y., Tsujii, J.: Boosting precision and recall of dictionary-based protein name recognition. In: Proceedings of the ACL 2003 Workshop on Natural Language Processing in Biomedicine, vol. 13, pp. 41–48. Association for Computational Linguistics (2003)
172. Volz, J., Bizer, C., Gaedke, M., Kobilarov, G.: Silk-a link discovery framework for the web of data. LDOW **538** (2009). http://ceur-ws.org/Vol-538/
173. Voutilainen, A.: Part-of-speech tagging. In: The Oxford Handbook of Computational Linguistics, pp. 219–232. Oxford University Press, Oxford (2003)
174. Vrandečić, D., Krötzsch, M.: Wikidata: a free collaborative knowledgebase. Commun. ACM **57**(10), 78–85 (2014)
175. Wactlar, H.D.: New directions in video information extraction and summarization. In: Proceedings of the 10th DELOS Workshop, Sanorini, pp. 24–25 (1999)
176. Wang, Q., Mao, Z., Wang, B., Guo, L.: Knowledge graph embedding: a survey of approaches and applications. IEEE Trans. Knowl. Data Eng. **29**(12), 2724–2743 (2017)
177. Wang, Q., Wang, B., Guo, L., et al.: Knowledge base completion using embeddings and rules. In: IJCAI, pp. 1859–1866 (2015)
178. Wang, Z., Zhang, J., Feng, J., Chen, Z.: Knowledge graph and text jointly embedding. In: Proceedings of the 2014 Conference on Empirical Methods in Natural Language Processing (EMNLP), pp. 1591–1601 (2014)
179. Wang, Z., Zhang, J., Feng, J., Chen, Z.: Knowledge graph embedding by translating on hyperplanes. In: AAAI, vol. 14, pp. 1112–1119 (2014)
180. Ward, M.D., Beger, A., Cutler, J., Dickenson, M., Dorff, C., Radford, B.: Comparing gdelt and icews event data. Analysis **21**(1), 267–97 (2013)
181. Wei, Z., Zhao, J., Liu, K., Qi, Z., Sun, Z., Tian, G.: Large-scale knowledge base completion: inferring via grounding network sampling over selected instances. In: Proceedings of the 24th ACM International on Conference on Information and Knowledge Management, pp. 1331–1340. ACM (2015)
182. Weld, D.S., Bansal, G.: Intelligible artificial intelligence. arXiv preprint arXiv:1803.04263 (2018)
183. Winkler, W.E.: Improved Decision Rules in the Fellegi-Sunter Model of Record Linkage, pp. 274–279. Bureau of the Census (1993)

184. Winkler, W.E.: Methods for record linkage and Bayesian networks. Technical report, Statistical Research Division, US Census Bureau, Washington, DC (2002)
185. Xie, R., Liu, Z., Jia, J., Luan, H., Sun, M.: Representation learning of knowledge graphs with entity descriptions. In: AAAI, pp. 2659–2665 (2016)
186. Xie, R., Liu, Z., Sun, M.: Representation learning of knowledge graphs with hierarchical types. In: IJCAI, pp. 2965–2971 (2016)
187. Yates, A., Cafarella, M., Banko, M., Etzioni, O., Broadhead, M., Soderland, S.: Textrunner: open information extraction on the web. In: Proceedings of Human Language Technologies: The Annual Conference of the North American Chapter of the Association for Computational Linguistics: Demonstrations, pp. 25–26. Association for Computational Linguistics (2007)
188. Yu, S., Bai, S., Wu, P.: Description of the kent ridge digital labs system used for MUC-7. In: Seventh Message Understanding Conference (MUC-7): Proceedings of a Conference Held in Fairfax (1998)
189. Zelenko, D., Aone, C., Richardella, A.: Kernel methods for relation extraction. J. Mach. Learn. Res. **3**, 1083–1106 (2003)
190. Zheng, S., Hao, Y., Lu, D., Bao, H., Xu, J., Hao, H., Xu, B.: Joint entity and relation extraction based on a hybrid neural network. Neurocomputing **257**, 59–66 (2017)
191. Zhou, G., Su, J.: Named entity recognition using an HMM-based chunk tagger. In: Proceedings of the 40th Annual Meeting on Association for Computational Linguistics, pp. 473–480. Association for Computational Linguistics (2002)

Index

A

Ablation-style evaluation, 72
Abstract properties, 16
Alexa, 87
Alias, 20
Analogy, 21, 61
Arguments, 6, 23
Assertion, 2
AUC-PR, 73
Automated ER solutions, 52
Automatically generated discriminator phrases,
   16
Automatic content extraction (ACE), 22
Automation-scalability tradeoff, 54

B

Background corpus, 17
Background knowledge, 35
Bags of words, 18
Baseline statistical learning, 13
Baseline system, 13
Bayesian optimality, 52
BBN ACCENT, 37
Big Data, 1
Bi-grams, 18
Bilateral pairs, 38
Blocking, 39
   key, 40
      learning, 43
      quality, 43
      value, 40
   method, 40
   without supervision, 56

Block purging, 40
Bootstrapping, 14

C

Candidate entity, 16
Candidate set, 44
   reduction, 48
Canopies, 42
Capitalization feature, 16
Case citations, 14
CBOW, 21
Character makeup, 17
Character n-gram, 17
Chunk, 18
Clerical review, 52
Clustering, 15
COALA, 53
Common corpus, 21
Common nouns, 19
Commonsense reasoning, 1
Conditional random fields, 10
Context, 20, 21, 61
Contextual clues, 14
Coreference resolution, 12, 20
Cortana, 87
Cosine similarity, 18
Cross-document coreference resolution, 33
Cue words, 19

D

Data ecosystems, 36
Data skew, 36, 40, 43

DBpedia, 18, 78, 80
Dedoop, 54
Deduplication, 34
Deep neural networks, 16
DeepWalk, 62
Derivational suffixes, 19
Diacritics, 19
Dictionary, 18
Disambiguation, 19
    rules, 13
Discriminative features, 13
Disjunctive normal form, 43
Distributional similarity, 14
DKRL, 69
Document-centric features, 19
Document and Corpus Features, 19
Domain adaptation, 37
Domain engineering, 45
Domain-specific KGC, 10
Downstream NLP, 61
Dublin Core, 79
Duplicated coverage, 48

E
Edit-distance, 19
Energy-based embedding model, 63
Entity classification, 73
Entity Resolution (ER), 6, 33
    automation, 36
    heterogeneity, 36
    related work, 51
    scalability, 36
Entity typing, 18
EUCLID, 53
Events, 6
    domain, 4
    extraction, 11, 24
Evolution of ER, 45
Exhaustive set, 49
Expectation maximization, 52
Extending the two-step workflow, 51
External lexicon, 18

F
Feature-based blocking, 42
Features, 16
    engineering, 16, 20, 61
    function, 17
    library, 45
    vector, 45
Fellegi-Sunter method, 44, 52
Finer-grained typing, 18

First-order entities, 6
Fixed-size window, 15
F-measure, 50
Freebase, 3
Fuzzy matching, 19

G
Gazetteer, 18
General AI, 1
Generalization, 13, 14, 66
    capabilities, 16
Genetic ER algorithms, 53
Geopolitical events, 37
GKV, 82
Global data space, 76
Global loss function, 63
Glove, 21
Google Knowledge Graph, 1
Google Knowledge Vault, 82
Google news corpus, 21
Graph-aware embedding models, 71
Graph embeddings, 38
Graph priors, 82
Greedy algorithms, 44
Ground-truth, 44

H
Hadoop, 41
Head entity, 2, 62
Heuristics, 45
Hierarchical graphical models, 52
HTML, 11
Human in the loop, 14
Hungarian algorithm, 47
Hyperparameter tuning, 66
Hyponyms/hypernyms, 15

I
I.i.d, 44
In-KG applications, 72
Inflectional, 19
Infobox, 80
Information extraction, 9
Information retrieval, 15
Instance-based blocking, 42
Internationalized resource identifiers (IRI), 5

J
Jaro-Winkler, 19
Joint IE, 25

Joint modeling, 24
Joint text-video extraction, 10
JSON-LD, 86

K
Knowledge bases, 3, 4
Knowledge graph
  adoption, 75
  completion, 59
  ecosystems, 75
  identification, 59
Knowledge graph embeddings (KGEs), 60
  applications, 72
Knowledge panels, 1
Knowledge repository, 82
Knowledge sub-model, 68
Knowledge Vault, 82

L
Labeled training data, 10
Latent Dirichlet allocation (LDA), 52, 61
Lemmatized, 19
Less well-known entities, 19
Lexical resources, 15
Lexicon, 18
LIMES, 54
Linguistic patterns, 14
Linked Data, 76
  principles, 77
  technology stack, 78
Linking Open Data (LOD), 77
Link prediction, 72
Link specification function, 38
List, 18
List-lookup features, 18
Literal denotation, 20
Literals, 9
Locality sensitive hashing (LSH), 54
Long short-term memory (LSTM), 16
Low-dimensional, 20

M
Machine learning effectiveness, 20
MapReduce, 41, 54
Markup, 9
Mean rank, 73
Mean reciprocal rank, 73
Measuring blocking performance, 48
Merge purge, 41
Message Understanding Conferences (MUCs),
  10

Meta-ability, 37
Metadata, 20
Meta-information, 19
Metaphone, 19
Metonymy, 20
Microformats, 86
Minimal supervision, 53
Monge-elkan, 47
Morphological feature, 17
MUC-6 training, 13
Multi-lingual, 17
Multi-pass sorted neighborhood, 42
Multi-relational graph, 62
Multi-word expression, 19
Mutual bootstrapping, 14

N
Named entity detection, 18
Named entity recognition (NER), 11, 12
N-ary, 22
Natural languages, 9
NLP-centric IE, 11
Nominal feature, 16
Normalizing, 19
NTN, 68
NYSIIS, 46

O
Ontological information, 67
Ontology matching, 56
OOV, 21
OpenCyc, 86
Open IE, 10, 13
Open Knowledge Network (OKN), 86
Out-of-KG, 69
Out-of-KG applications, 74

P
Pairs completeness, 49
Pairs quality, 49
Patient linking, 34
Pattern features, 18
Pattern generalization, 14
PC-RR tradeoff, 49
PFM, 53
Phonetic algorithm, 19
Phonetic code, 19
PMI-IR, 15
Pointwise mutual information, 15
Post-processing steps, 51
Precision, 15, 50, 73

Probabilistic soft logic (PSL), 66
Product matching, 37
Products and e-commerce, 5
Projection matrix, 65
Property heterogeneity, 37
Property matching, 51
Publication, 4, 39

Q
Quadratic complexity, 38
Quantitative feature, 16

R
Rare named entities, 15
RAVEN, 55
RDF, 62
RDFS, 79
Real-world ER, 46
Recall, 50, 73
Recall-friendly, 51
Receiver operating characteristic (ROC), 50
Red-blue set covering, 44
Reduction ratio, 49
Reference, 19
Relation-specific hyperplanes, 64
Relational features, 20
Relational information, 53
Relation Extraction (RE), 11, 22
Relation path, 71
Relationship, 2
Representation learning, 16, 38, 61
RoadRunner, 28
Robust, 16
Rule base, 45
Rules and KGEs, 69

S
Same semantic class, 14
Schema.org, 84
Schema heterogeneity, 51
Second-order entities, 6
Seed entity, 14
Segmented noun phrases, 16
Selecting the unlabeled data, 15
Semantic class, 21
Semantic dependence, 61
Semantic regularity, 60
Semantic relation, 22
Semantic tagging, 20
Semantic Web, 3, 33, 62, 76
Semi-supervised learning, 14

Sequence-labeling, 16
SILK, 54
Similarity, 44
    of context, 15
Site generation, 28
Skip-gram, 21
SKOS, 79
Sliding window, 41
Slot fillers, 9
Sorted neighborhood, 41
Soundex, 19
SPARQL, 77
Special character, 17
Spelling errors, 18
SSE, 67
Stable marriage, 53
Stemmed, 19
String similarities, 19
Structural homogeneity, 38
Structured data, 1
Supervised machine learning, 10, 13
Swoosh, 54
Synonyms, 14
Synset, 15
Syntactic units, 16

T
Tail entity, 2, 62
Temporal information and KGEs, 70
Terrorist attack events, 24
Textual description incorporation into KGE
    model, 68
Tf-idf, 18, 20, 46, 52, 61
Thresholded edit-distance, 19
Time-aware embedding, 70
TKRL, 67
Token-based, 18
Traditional blocking, 40
Trans*, 64
TransE, 63
TransE extensions, 64
Transfer learning, 53
TransH, 64
Translations, 63
TransR, 64
Tri-grams, 18
Triple, 2
Triple classification, 73
Triplify, 10
Twitter, 14
Two-layer similarity feature, 46
Two-step framework, 38
Type-specific projection matrix, 67

Type heterogeneity, 36
Type hierarchy, 37
Type matching, 51
Type signatures, 15

U
Underlying ontology, 4
Unlabeled corpora, 38
Unlabeled corpus, 15
Unsupervised learning, 14
Unsupervised machine learning, 15
URL, 77

V
Vector space, 61
Viable linking candidates, 35
Vocabulary transfer, 13

W
Weakly supervised, 14
Web IE, 11
Web of data, 75
Web of things, 76
Web queries, 15
Wikidata, 3, 86
Wikipedia, 3, 21
Wikipedia anchors, 69
Word-level features, 17
Word2vec, 21, 61
Word embeddings, 20
WordNet, 15
Word polysemy, 18

Y
YAGO, 18, 67

Printed in the United States
By Bookmasters